Lecture Notes in Bioengineering

More information about this series at http://www.springer.com/series/11564

Amit Gefen · Daphne Weihs
Editors

Computer Methods in Biomechanics and Biomedical Engineering

Proceedings of the 14th International Symposium CMBBE, Tel Aviv, Israel, 2016

Springer

Editors
Amit Gefen
Department of Biomedical Engineering
Tel Aviv University
Tel Aviv
Israel

Daphne Weihs
Faculty of Biomedical Engineering
Technion - Israel Institute of Technology
Haifa
Israel

ISSN 2195-271X ISSN 2195-2728 (electronic)
Lecture Notes in Bioengineering
ISBN 978-3-319-59763-8 ISBN 978-3-319-59764-5 (eBook)
DOI 10.1007/978-3-319-59764-5

Library of Congress Control Number: 2017943207

Printed on acid-free paper

This Springer imprint is published by Springer Nature
The registered company is Springer International Publishing AG
The registered company address is: Gewerbestrasse 11, 6330 Cham, Switzerland

Contents

Computational Fluid Dynamics for Intracranial Aneurysm Rupture Prediction and Post-treatment Hemodynamic Analysis

George Albert Einstein, Srinivasan Aishwarya, V. Sreeja and S. Nandhini

Abstract Intracranial aneurysm is a cerebrovascular disorder that weakens the intima of the cerebral artery or vein causing a local dilation of the blood vessel after which the blood vessel becomes thin and ruptures without warning. The resultant bleeding into the space around the brain is called subarachnoid hemorrhage (SAH). In this context, there is an urgent need for early aneurysm rupture prediction that could save numerous human lives. Computational fluid dynamics (CFD) has proved to be potent enough for the prediction of intracranial aneurysm rupture. The computational analysis along with the expertise in numerical methods for the virtual prediction of complicated life-threatening medical disorders is essential to be focused for the betterment of society. In this research, we aimed to predict the rupture of intracranial aneurysm through different morphological parameters of the aneurysm and changes in hemodynamics obtained from CFD. The various treatment options for intracranial aneurysm involves endovascular clipping, coiling and both coiling and stenting. We demonstrated quantitatively and qualitatively the hemodynamic changes (velocity, pressure, and wall shear stress—WSS) in artery after stent implantation and coiling into the 3D reconstructed patient-specific artery model, taking into consideration the non-Newtonian characteristics of blood using finite element approach. We also performed a comparison between scenarios of aneurysm and normal artery without aneurysm, which showed considerably high WSS, pressure, and velocity values in the model with aneurysm. The possible treatment options for the case were also computationally analyzed with the fluid structure interaction (FSI) which showed stenting and coiling considerably reduced the WSS on the aneurysm dome. It was speculated that low WSS at the tip of the aneurysm and high WSS on the aneurysm dome is responsible for the rupture of aneurysm.

G. Albert Einstein (✉) · S. Aishwarya · V. Sreeja · S. Nandhini
Department of Biosciences, EinNel Technologies, Chennai, Tamil Nadu, India
e-mail: einstein@einnext.com

A. Gefen and D. Weihs (eds.), *Computer Methods in Biomechanics and Biomedical Engineering*, Lecture Notes in Bioengineering,
DOI 10.1007/978-3-319-59764-5_1

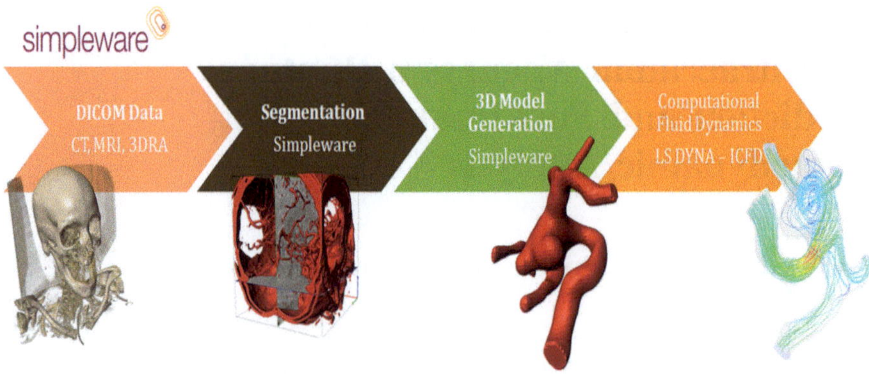

Fig. 1 Process workflow

Introduction

The incidence of intracranial aneurysm varies from 0.75 to 10.3% in India, out of which approximately 0.2–3% of the people suffering from intracranial aneurysm may suffer from subarachnoid hemorrhage per year (Rinkel et al. 1998). This intracranial aneurysm can occur in people of all ages but are most commonly detected in those ages of 35–60. Women are found most likely to get intracranial aneurysms than men with the ratio of 3:2.

In this context we find the importance of early aneurysm rupture prediction that could save numerous human lives. Computational fluid dynamics has proved to be potent enough for the prediction of intracranial aneurysm rupture.

EinNel Technologies envisage improving the quality of human lives in every feasible manner. In this regard we have started our new venture of biosciences comprising various arenas in biomedical and biotechnology. We focus on computational analysis with our expertise in numerical methods for the virtual prediction of complicated life-threatening medical disorders (Fig. 1).

Materials and Methods

Modeling of the Intracranial Aneurysm

3D Reconstruction Technique for Intracranial Aneurysm

The construction process of image-based artery model involves the following three major steps:

(1) Non-invasive Image acquisition,
(2) Imaging processing and
(3) Three-dimensional reconstruction to form voxel-based volumetric image representation.

We obtained DICOM data of aneurysm case from 3D Rotational Angiography of the right or left internal carotid artery with middle cerebral artery (MCA) aneurysms with Simpleware Scan IP v2.0.20 software for generation of 3D reconstructed images (Cebral et al. 2011). The complete blood vessel domain was isolated from the volume rendered model. The fine surface irregularities were removed while refining surface using smoothening in the scan IP software package, and the 3D surface was exported in stereo-lithography (STL).

Generation of Mesh Model

STL files for stenting models were defined with finite element meshes in meshing tool. The mesh refinement and quality checks maintaining uniform triangular element lengths of the range 0.06–0.15 mm for was performed (Fig. 2).

Three cases of simulation is considered according to the treatment options namely,

1. Without coil and stent
2. With coil and no stent
3. With coil and stent
4. Without aneurysm

Therefore, accordingly these models are modeled for simulation.

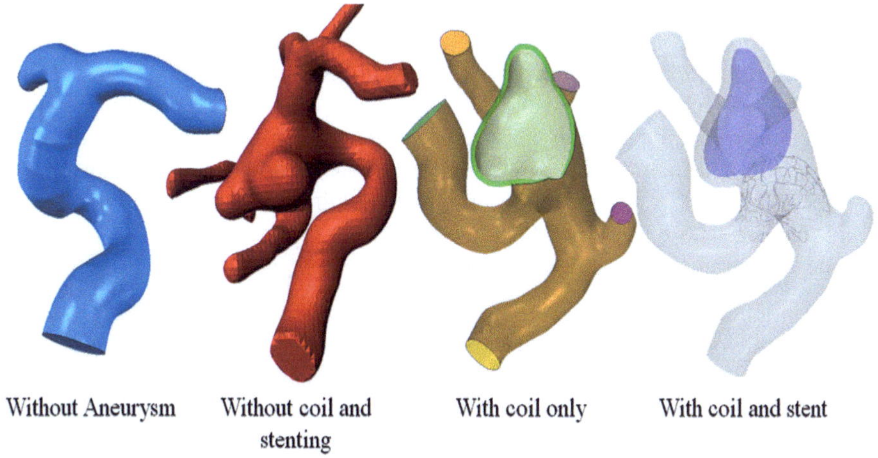

Fig. 2 MCA models with different treatment conditions

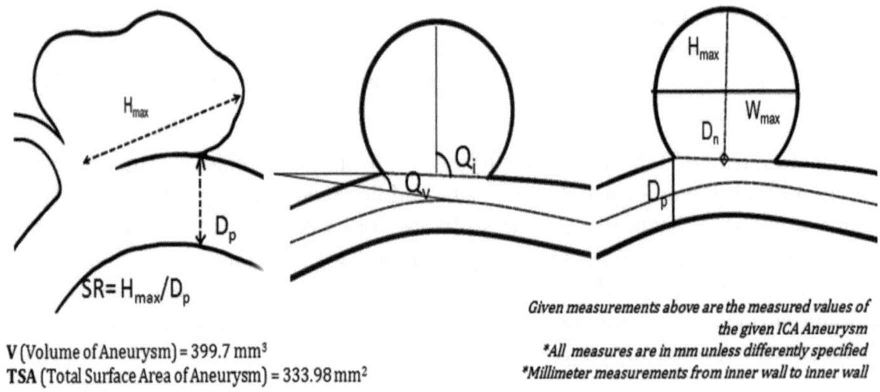

Fig. 3 Morphological measurements

Morphological Study

The morphological study on the aneurysm was an attempt to identify the level of rupture risk which can be predicted only on the basis of its structure, area, and volume. Some of the main parameters that are considered are size ratio (SR), inclination neck angle (Q_i) and the size (S) which is the total size of the aneurysm bleb. SR and S showed more significance to rupture prediction on unmanageable aneurysm (Fig. 3).

The given case was found to be right internal carotid artery (ICA) aneurysm. These inferences stated below are for predicting rupture using morphology of aneurysm by comparing the obtained morphological values of the given ICA case with literature (Tremmel et al. 2009):

- Size ratio of given ICA aneurysm—2.34 (SR has highest significance for IA rupture risk, 77% of ruptured aneurysms had SRs greater than 2.05),
- Inclination neck angle of given ICA aneurysm—107.7° (80% of all ruptured IA have aneurysm inclination neck angle greater than 100° IA with daughter aneurysm and inclination angle >100° has more significance to rupture),
- Size of given ICA aneurysm—8.9 mm (size with 10 mm and more had risk of IA rupture with 85.6% of SAH occurrence).
- The 3D parameters found are, namely,
  - Undulation index (UI) is 0.3
  - Ellipticity index (EI) is 0.07
  - Non-sphericity index (NSI) is 0.34 are significant to intracranial aneurysm rupture.
- In reference to these morphological values SR, inclination angle, EI and NSI converging with the ruptured values we can predict that the given ICA aneurysm is probably ruptured or more prone to rupture.

- The above prediction is purely based on morphological parameters of given ICA aneurysm and the computational prediction will be presented in the future.

Numerical Simulation

CFD simulations were performed in LS-DYNA ICFD solver which solves the three-dimensional incompressible Navier–Stokes equations in an arbitrary domain.

Blood was modeled as incompressible Newtonian fluid with a density of 1060 kg/m^3 and viscosity 0.0035 Pa. A linear inlet velocity of 0.8 m/s was applied considering the peak systolic blood velocity. An outlet pressure of 0 Pa was maintained and a no-slip condition was used at the walls. All calculations were performed in steady state condition and K-epsilon Turbulence model was used for simulation with standard K-epsilon constants values, viz., $C_\mu = 0.09$, $Ce_1 = 1.44$, $Ce_2 = 1.92$, $C_k = 0.6$, $\sigma_k = 1$, $\sigma_e = 1.3$.

A time step of 4E−5s was used for all the simulations. The approximate calculation time for one case when run with a 4-core license in a 12-core Xeon Processor, 3.4 GHz, 32 GB RAM was 72 h and the convergence criteria chosen was, residuals <1E−6.

The wall shear stress (WSS) distribution for all MCA aneurysms was analyzed. The velocity and pressure profiles were also obtained for all the scenarios. For an aneurysm the diameter of the inlet of aneurysm and the maximum height of the aneurismal sac were measured from the 3D model. From these values, the aspect ratio was calculated by dividing the height by the diameter obtained. The rupture status predicted by us is purely based on WSS, velocity, and pressure of all the aneurismal cases.

In order to understand the variation in pressure dynamics in the case of the normal artery and artery after developing an aneurysm, we modified the model removing the aneurysm and performed CFD. There is a considerable decrease in the overall WSS ($\sim$32 Pa) for the no aneurysm scenario (shown in Fig. 4d). The other scenario is placing a porous solid resembling the coil placed in actual treatment and obtained the hemodynamic results (shown in Fig. 4b). After placing a porous coil in the aneurysm the flow into the bleb visibly reduced. The pressure and WSS values increased compared to the no-coil and no-stent aneurysm model. This can be attributed to the turbulence generated at the neck which is considerably wide to let enough flow hit the packed coil creating eddies.

So we concluded that coiling alone cannot be a feasible solution. The results clearly shows the treatment of coiling and stenting has reduced the WSS over the aneurysm (shown in Fig. 4c). There is a considerable reduction in flow into the aneurysm after coiling. Moreover the velocity magnitude at the narrowed zones of artery has decreased. When the stress parallel to the wall is seemingly increasing then there occurs to be increasing risk of aneurysm rupturing. The model without coil and stent, i.e., the pre-surgical model had the highest WSS ($\sim$42 Pa). We can recall that the model without aneurysm which is considered as the baseline had most reduced WSS. Relating with all these inferences we came up with an

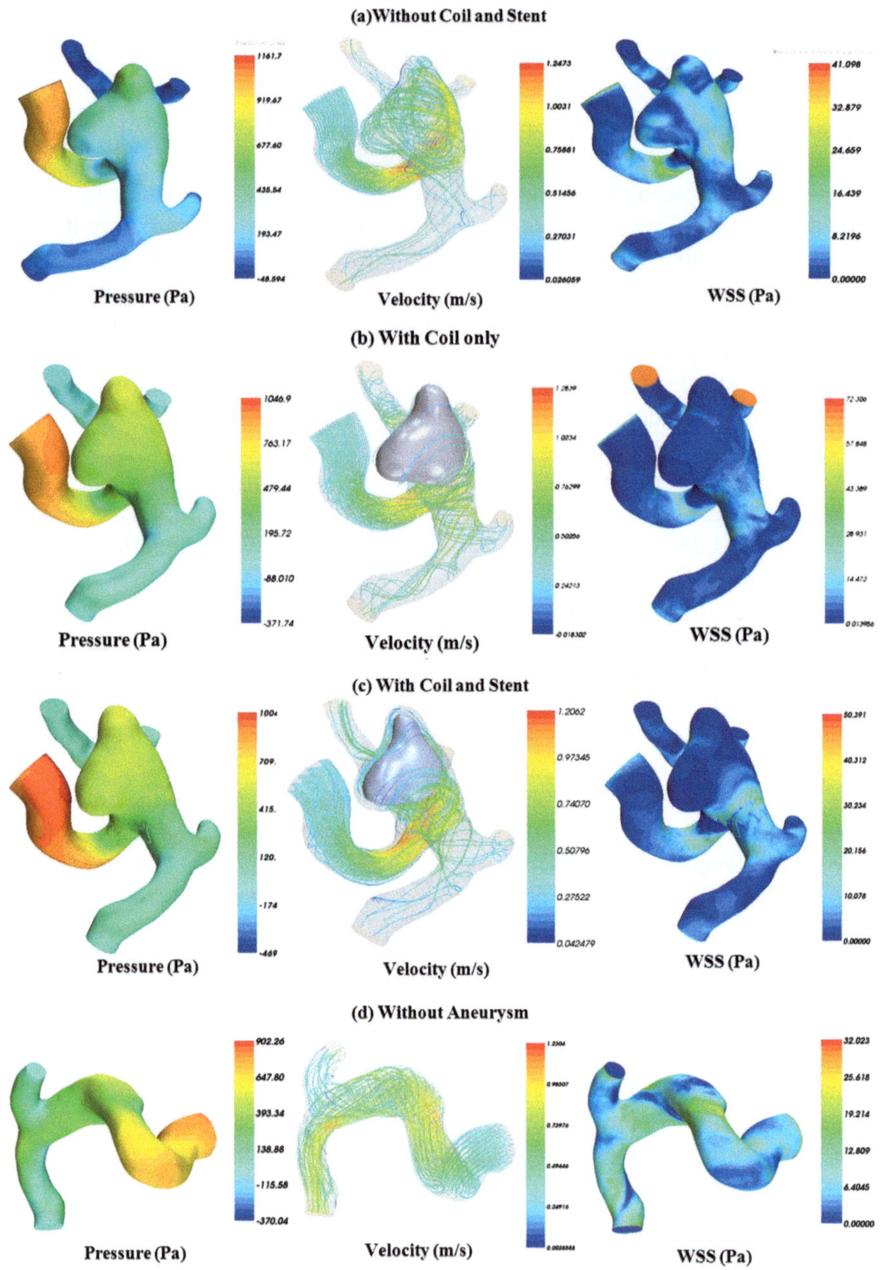

Fig. 4 Hemodynamics for different scenarios

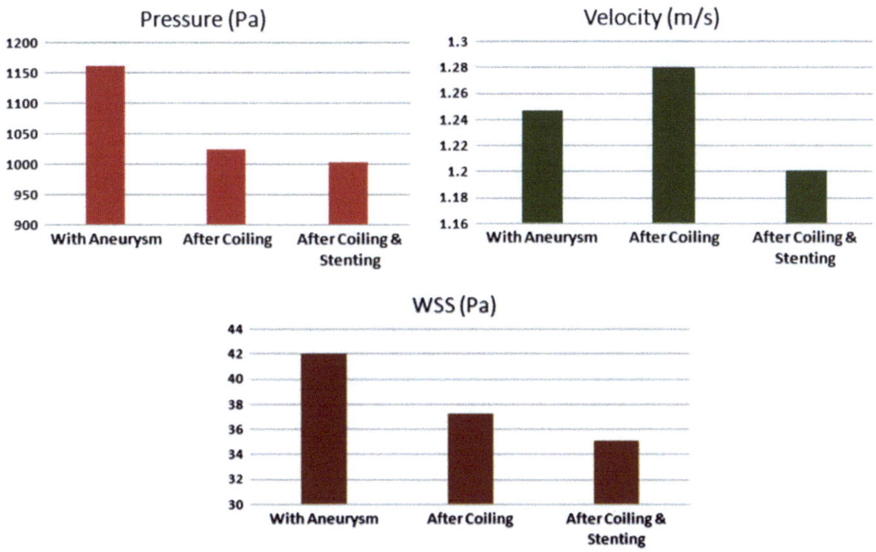

Hemodynamic Index	With Aneurysm	Without Aneurysm	Aneurysm with Coil	Aneurysm with Coil and stent
Pressure (Pa)	1161.1	902.26	1025	1004.6
Velocity (m/s)	1.247	1.230	1.28	1.201
WSS on Aneurysm	42	32.03	37.3	35.1

Fig. 5 Overall result comparison

observation that after installing a coil and a stent the risk of aneurysm rupture gets reduced. And so analyses confirm that coiling and stenting could considerably reduce the hemodynamic indices values compared to that of untreated aneurysm (refer Fig. 5).

The fluid–structure interaction (FSI) analysis was done in a single platform for solving structural and fluid problems. The stented artery model was analyzed ICFD-FSI capability in LS-DYNA solver. The artery is modeled with biological tissue material and the grafts such as the stent and coil is modeled with shape memory alloy material NiTinol. The fluid medium was considered to be non-Newtonian using the power law and Carreau models (Figs. 6 and 7).

The main aim of introducing the fluid and the structure into a single process is to analyze the interaction between the artery, blood and the stent. The ICFD-FSI in LS-DYNA helps to identify the hemodynamic changes before and after stenting.

The displacement of artery and stent due to the force created from the blood flow is measured. The stent had more displacement than the artery due to the fact that stent is more interacted with the blood flow than the artery. We observed that the

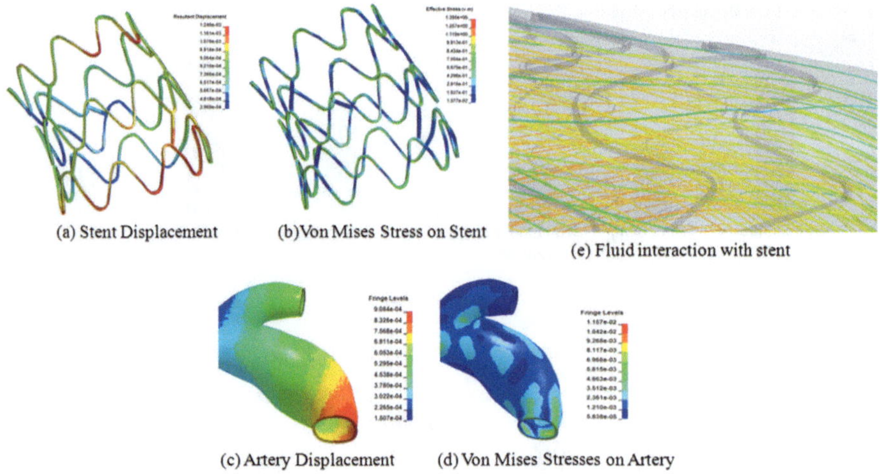

Fig. 6 FSI results on stent and artery

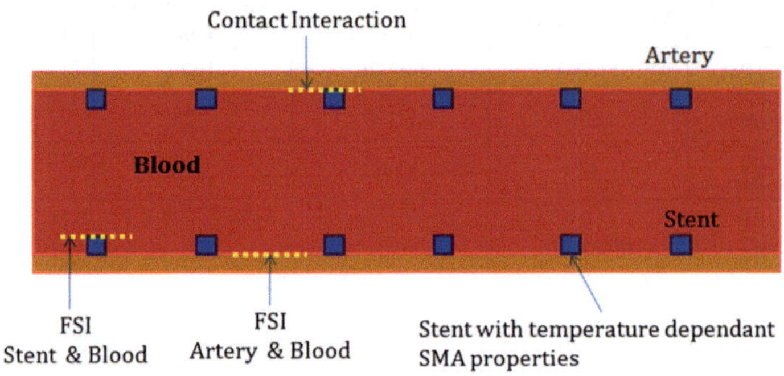

Fig. 7 Blood, artery and stent interactions

displacement is more at the stent edges faced towards the blood flow and comparatively less displacement on stent edges faced towards the artery side. Hence we can accept that the stent tends to migrate in accordance with the flow direction and force.

Discussion

The aneurysm rupture risk prediction continues to be a debate, it occurs to be that the morphological parameters stratification of the intracranial aneurysms has great significance in rupture prediction rather than considering the overall size of the aneurysm bleb alone (Rahman et al. 2010). The computational fluid dynamics (CFD) of the hemodynamics inside the aneurysm affected blood vessel had brought more insights about the flow and rupture prediction. The CFD is proved to be the potent method to assess aneurysm flow non-invasively (Tremmel et al. 2009). Aneurysms with increased neck diameter in a curved vessel tend to have more hemodynamic stresses likewise our case in this study (Cebral et al. 2011).

Predicting the flow dynamics in large and prominent arteries is easier where as small tortuous arteries like intracranial arteries is tedious. The currently available techniques in the medical imaging environment like phase contrast MRV, which screens arteries like aorta, carotid arteries, etc. are said to be more prominent and handy to manage unlike the IAs. Hence to study the hemodynamic indices in these IAs we can consider that computational analysis is the only cogent method for aneurysm rupture prediction (Shojima et al. 2004).

The WSS is the ultimate parameter to decipher the rupture status and its risk (Goubergrits et al. 2012). It is contemplated that areas of steep transitions in high and low WSS subvert the aneurysm vessel wall and thus leading to rupture. The aneurysm geometry changes with the event of rupture and the flow patterns. The hemodynamic flow from the parent artery which hits the gateway of the aneurysm neck in a particular direction has more significance towards the rupture spot prediction (sys-risk of rupture). In our study, the flow patterns entering the neck of aneurysm and the flow circulations inside the aneurysm were analyzed even in the presence of coil and stent scenarios in order to get an idea of the treatment options in pre-surgery stage (Cebral et al. 2005).

Conclusion

The intracranial aneurysm (IA) was analyzed in two major methods, i.e., the morphological assessment and the CFD assessment. Using these methods we calculated that the given internal carotid artery (ICA) aneurysm case was highly prone to rupture and there is an immediate requirement for coiling and stenting. After performing different scenarios, we found that the untreated aneurysm had the highest WSS, velocity, and pressure, whereas the model with stent and coil showed considerably decreased WSS, velocity, and pressure. The WSS parameter is considered to be the significant factor for rupture risk. The WSS values for the coil and stent model matched approximately with the baseline model without aneurysm. We came to a conclusion that the CFD method helps to analyze the therapeutic conditions in the pre-surgical stage of the patient without affecting them. This computational prediction enables the surgeons to come up with ideas to deal with unmanageable intracranial aneurysm (IAs).

Acknowledgements We would like to thank the doctors, who guided this work, Dr. Prithika Chary MD, DM, Ph.D., M.Ch., Senior Consultant Neurologist, and Neurosurgeon, Kauvery Hospital. Dr. Satheesh Grahadurai MBBS, DMRD, M.Sc., FINR Consultant Interventional Neuro Radiologist Kauvery Hospital. Dr. Iyappan Ponnuswamy MD, FRCR Consultant Radiologist, Kauvery Hospital.

References

Cebral JR, Castro MA, Burgess JE, Pergolizzi RS, Sheridan MJ, Putman CM (2005) Characterization of cerebral aneurysms for assessing risk of rupture by using patient-specific computational hemodynamics models. AJNR Am J Neuroradiol 26:2550–2559

Cebral JR, Mut F, Weir J, Putman C (2011) Quantitative characterization of the hemodynamic environment in ruptured and unruptured brain aneurysms. AJNR Am J Neuroradiol 32:145–151

Goubergrits L, Schaller J, Kertzscher U, van den Bruck N, Poethkow K, Ch Petz H-Ch, Hege A Spuler (2012) Statistical wall shear stress maps of ruptured and unruptured middle cerebral artery aneurysms. J R Soc Interface 9:677–688

Juvela S, Poussa K, Porras M (2001) Factors affecting formation and growth of intracranial aneurysms: a long-term follow-up study. Stroke 32:485–491

Kataoka K, Taneda M, Asai T, Kinoshita A, Ito M, Kuroda R (1999) Structural fragility and inflammatory response of ruptured cerebral aneurysms a comparative study between ruptured and unruptured cerebral aneurysms. Stroke 30:1396–1401

Miura SY, Ishida F, Umeda Y, Tanemura H, Suzuki H, Matsushima S, Shimosaka S, Taki W (2013) Low wall shear stress is independently associated with the rupture status of middle cerebral artery aneurysms. Stroke 44:519–521

Rahman M, Smietana J, Hauck E, Hoh B, Hopkins N, Siddiqui A, Levy EI, Meng H, Mocco J (2010) Size ratio correlates with intracranial aneurysm rupture status, a prospective study. Stroke 41(5):916–920

Rinkel GJE, Djibuti M, Algra A, van Gijn J (1998) Prevalence and risk of rupture of intracranial aneurysms, a systematic review. Stroke 29:251–256

Shojima M, Oshima M, Takagi K, Torii R, Hayakawa M, Katada K, Morita A, Kirino T (2004) Magnitude and role of wall shear stress on cerebral aneurysm, computational fluid dynamic study of 20 middle cerebral artery aneurysms. Stroke 35:2500–2505

Tremmel M, Dhar S, Levy EI, Mocco J, Meng H (2009) Influence of intracranial aneurysm-to-parent vessel size ratio on hemodynamics and implication for rupture: results from a virtual experimental study. Neurosurgery 64(4):622–663

Biomechanical Evaluation of Pre- and Post-bilateral Sagittal Split Mandibular Osteotomy on Three-Dimensional Models for Obstructive Sleep Apnea Using Finite Element Analysis

Srinivasan Aishwarya, George Albert Einstein, S. Nandhini and K. M. Vinod

Abstract Obstructive sleep apnea syndrome (OSA) is the hindrance of upper airway during sleep, associated with curtailment in blood oxygen saturation. It is characterized by subdual flow of oxygen to vital organs causing irregular heart rhythms. One of the triumphant surgeries to treat OSA is maxillo-mandibular advancement (MMA) which is found to be 90% successful for OSA patients (Phee 2015). In MMA, the lower jaw and the mid-face are progressed to augment the posterior airway space facilitating trouble-free breathing. In this research, we attempted to contemplate the von Mises stresses due to mastication in normal and osteotomed 3D models and identify the maximum stress that can be tolerated by the mandible using finite element analysis (FEA). FEA has been extensively used to solve complex problems in dentistry and researchers have found a high correlation between FEA simulation results and in vitro measurements for mandibular specimens (Erkmen et al. 2005). The location of screws and miniplate fixation in the 3D osteotomed models was determined by Champy's lines in order to ensure stable fixation (Erkmen et al. 2005). We first evaluated the extent of movement of the posterior airway space that is mandatory for the OSA patients to breathe normally. It was evident that the airway constriction was corrected in the upper respiratory tract by the advancement of the mandible. The von Mises stress and displacement in the mandible before and after MMA by applying three different loads, incisal, contralateral compressive molar loads, and one-sided molar loads were analyzed to rule out the fixation and orthognathic issues. The stress distributions during mastication were furthermore compared for mandibular osteotomy models with two distinct lengths of advancement. In addition, the deflection by virtue of mastication on molars, incisors, and canines was also assessed. In line with the above-mentioned evaluation, we performed the computational fluid dynamics

S. Aishwarya · G. Albert Einstein (✉) · S. Nandhini · K.M. Vinod
Department of Biosciences, EinNel Technologies, Chennai, Tamil Nadu, India
e-mail: einstein@einnext.com

A. Gefen and D. Weihs (eds.), *Computer Methods in Biomechanics and Biomedical Engineering*, Lecture Notes in Bioengineering,
DOI 10.1007/978-3-319-59764-5_2

(CFD) analysis of the upper airway model with the pre- and post-surgical conditions to predict the airflow dynamics accordingly.

State of the Art

Obstructive sleep apnea (OSA) is the most prevalent type of sleep apnea, caused by partial or complete blockage of upper respiratory tract during sleep. It is indicated by continual episodes of shallow or paused breathing during sleep and is usually associated with the reduction in blood oxygen saturation.

According to the statistical study by International Society of Aesthetic Plastic Surgery (ISAPS) of 2015, for around 0.4 million face and head procedures are performed globally. There are several surgical treatments to modify airway anatomy that, known as sleep surgery, are varied and must be tailored to the specific airway obstruction needs of a patient. For those obstructive sleep apnea sufferers unable or unwilling to comply with front line treatment, a properly selected surgical intervention will be the result of considering an individual's specific anatomy and physiology, personal preference and disease severity (Sato et al. 2012). There is little randomized clinical trial evidence for all types of sleep surgery. One of the triumphant surgeries to treat OSA is maxillo-mandibular advancement (MMA), a type of orthognathic surgery which is found to be 90% successful for OSA patients (Phee 2015). Further this surgery offers an equally efficacious alternative in mild to moderate OSA patients who are not compliant or refuse CPAP therapy. The lower jaw and the mid-face are progressed in MMA to augment the posterior airway space and facilitate trouble-free breathing. The most common orthognathic surgical procedure is bilateral sagittal split osteotomy (BSSO).

Trauner and Obwegeser in 1957 gave the first description of this orthognathic technique Takahashi et al. (2010). Since then, Dal Pont introduced many modifications with the aim of improving surgical convenience, minimizing morbidity, and maximizing procedural stability, which is generally recognized that the buccal osteotomy cut of the Obwegeser–Dal Pont method is positioned more anteriorly than that of the Obwegeser method (Takahashi et al. 2010), thus increasing the cancellous bone contact.

The proximal and distal segments after the procedure of BSSO can be stabilized by rigid internal fixation for speedy bone healing triggering promptly postoperative mandibular function, and decreasing the amount of relapse (Takahashi et al. 2010). For an ideal stable osteotomy design, there are certain primary factors affecting the stability. Not only the miniplate orientation and shape and the location of the osteotomy cut, but also the location of the miniplates (superior, middle, or inferior) was considered to be the main parameter by using finite element analysis (FEA) (Champy et al. 1978). Champy et al. (1978) ascertained "the ideal line of osteosynthesis in the mandible" where miniplate fixation is the most stable.

The complex problems in dentistry can be solved using FEA. Vollmer et al. (2000) have found quite a high correlation between FEA simulation and in vitro measurements of mandibular specimens. FEA is thus an efficacious research tool which can endow clear-cut insight into the mandible's complex mechanical behavior, which is quite hard to assess by other means.

In this research, we attempt to compute the biomechanical behavior of the mandible and screw-miniplate system. We applied incisal and contralateral molar compressive loads to contemplate the deflection as well as the maximum von Mises stresses in the mandible before and after the surgery. Two models with distinct levels of advancement in the MMA were done and the same was compared and analyzed.

Materials and Methods

Modeling of the Mandible

The aim of modeling is to simulate the original scenario of BSSO. We have two different types of models; one is the normal mandible and the other is the osteotomed model of the mandible with fixation of screws and miniplates.

(i) *Volume rendering of the mandible*:

A stack of 2D computed tomography (CT) images with each image having 1 mm thickness were processed to render a 3D mandibular model.

(ii) *Levels of progression in the BSSO models*:

The various levels of advancement, viz., 5 and 7 mm in BSSO for the treatment of obstructive sleep apnea patients were modeled to evaluate the biomechanical behavior of the mandible post-surgery.

(iii) *Screws and miniplates modeling*:

Monocortical screws and miniplates are placed during osteotomy to withstand the loads exerted in the bone after surgery. As aforementioned, two pairs of monocortical plates are placed in the osteotomed model where six-screw-miniplates and two-screw-miniplates on each side of the mandible. The miniplates are approximately 40 mm in length and 1 mm in thickness. The screw with head dimensions of 2.5 mm width and 1 mm height, and its body dimensions of 2 mm width and 5 mm height was modeled. As per AO Foundation (www2.aofoundation.org), if the height of the screw goes beyond 6 mm, it might affect the tooth's roots. So the height of the screw is maintained at an optimum range (Fig. 1).

(iv) *Considerations for fixation of the implant*:

The oblique line is the line which passes just below the mental foramen. We looked upon this oblique line as reference for the osteotomy cut and constructed the mandibular advancement in the model. Fixing the screws and miniplates in the mandible should come hand-in-hand with experience of the surgeon as the

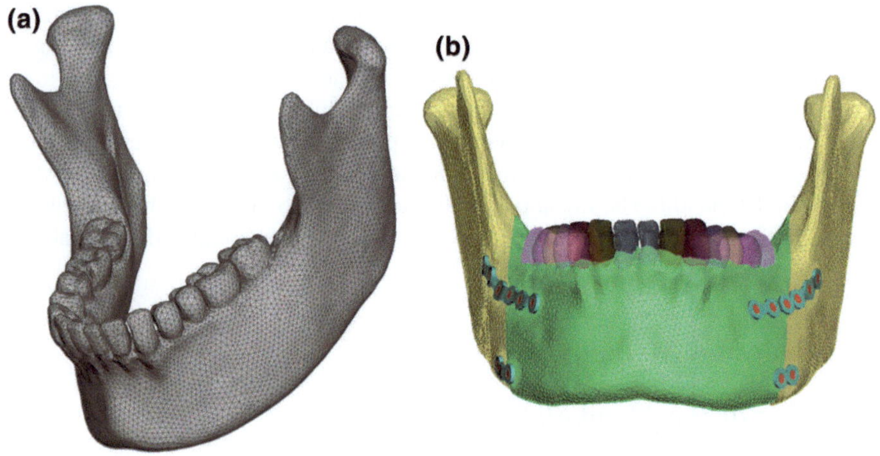

Fig. 1 **a** Normal mandible. **b** Osteotomed mandible

placement of the implant should not affect the roots of tooth, nerve, blood vessels, etc. We also have taken into account the ideal line of osteosynthesis, Champy's line before fixation of the implant on the osteotomed mandible (Champy et al. 1978).

(v) *Contouring of miniplates*:

Contouring is the process of twisting the miniplates with the help of two pliers at both ends to ensure matching of the underneath bone surface. Surgeons usually contour the miniplate according to the surface anatomy which is adjacent to the fraction line for a perfect and stable fixation. This ensures adaptation of the plate with the surface from one end to the other end. The intermediate bone surface which does not get along with the miniplate surface is excised during the surgery that is also followed during our modeling of the implant. We performed contouring through our in silico pre-processing methods.

Upper Airway path Modeling

The 3D model of upper pharyngeal airway is obtained from the same BSSO case. The airway region is segmented using threshold function with Simpleware Scan IP software. The model is meshed with approximately 0.7 million polyhedral volume elements. The fluid volume model is meshed with three boundary layers. Airway path extracted in this study extends from nostrils till the trachea and the thyroid gland. Two different airway paths were considered in this study: the pre-surgical airway path and the extended path. The extension is performed at the constriction from the oropharynx to laryngopharynx synching with the mandibular advancement (Fig. 2).

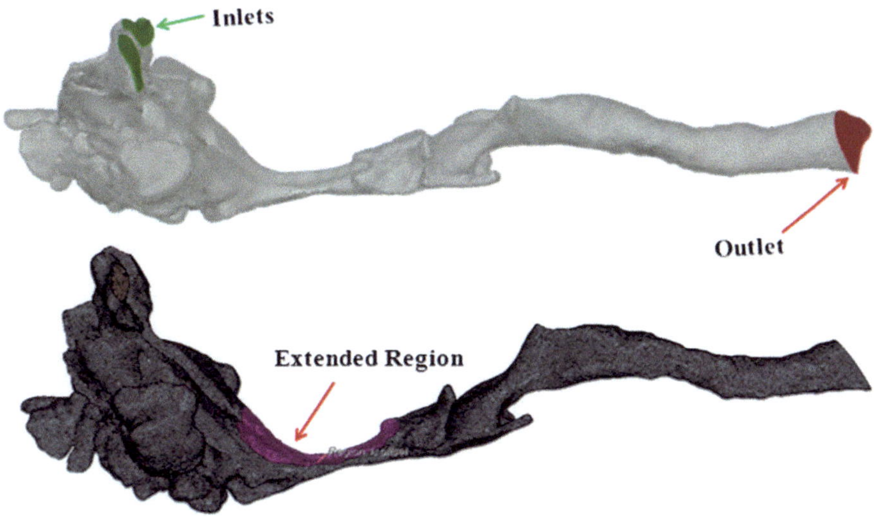

Fig. 2 Upper airway path

Materials used for the Modeling

(i) *Software tools utilized for modeling and analysis*:

The segmentation process of the region of interest (mandible) of DICOM images from CT scan was performed using Simpleware, ScanIP 7.0. The processed 3D image was exported in the stereolithography (STL) format. The rendered STL files of the mandible were imported into a finite element pre-processing tool for discretization for finite element meshing. The resulting model consisted of 0.4–0.5 million tetra elements in the model. The simulations and analysis were executed using Abaqus Implicit solver.

Boundary Conditions—BSSO

(i) *Constraints*:

The temporo-mandibular joint (TMJ) is bilaterally constrained in all degrees of freedom. The goal of constraining TMJ is to simulate the masticatory movement of the mandible only.

(ii) *Material properties*:

The material used for different types of teeth in the pre-surgical model is the dentin which is of more composition mimicking the teeth properties. We considered the properties of titanium for the screws and miniplates that are in the post-

Table 1 Material properties of various materials used

Material	Young's modulus (MPa)	Poisson ratio	Density (ton/mm^3)
Tooth (dentin)	1400	0.32	$2.65e^{-9}$
Mandible (cortical bone)	9000	0.3	$1.98e^{-9}$
Implant (titanium)	115,000	0.33	$4.43e^{-9}$

Table 2 Loads applied to incisors (Biswas et al. 2013)

Tooth	Incisal loads (N)	
	Right side	Left side
Central incisor	200	
Lateral incisor		
Canine		

Table 3 Loads applied to molars (Biswas et al. 2013)

Tooth	Contralateral molar loads (N)	
	Right and left side	One-sided
First premolar	300	
Second premolar		
First molar		
Second molar		

surgical model. The properties of various materials, viz., Young's modulus, Poisson ratio, and density are given (Table 1):

(iii) *Applied masticatory loads*:

The three different types of loads, i.e., incisal, contralateral molar loads, and one-sided contralateral molar loads are applied according to the anatomy of the tooth. Loads are applied perpendicular to the occlusal plane. The human mastication is a complex biomechanical process. During mastication, it is obvious that persons masticate harder substances with higher pressure and strength using the molars when compared to incisors (Tables 2 and 3).

Boundary Conditions—Airflow Dynamics

Airflow inside the human nasal cavity at rest and at dynamic state is taken under consideration. During normal breathing 15 l/min of air is inhaled; in this study the upper pharyngeal airway is simulated in normal breathing condition. The flow is kept steady throughout the simulation and air is modeled with K-epsilon turbulence model. Airflow inlet was modeled with mass flow rate inlet, outlet to be pressure outlet, and the wall to be with no-slip boundary condition (Table 4).

Table 4 Boundary conditions

Descriptions	Quantity	Air properties	
Space	Three dimensional		
Time	Steady state	Density	1.225 kg/m^3
Material	Air		
Flow	Segregated flow, constant density	Dynamic viscosity	1.85508E−5 Pa s
Model	Realizable K-epsilon turbulence		
Boundary conditions			
Inlet		Mass flow rate	
Outlet		Pressure outlet	
Wall		No-slip	

Results

To obtain the responses of mastication in mandible before and after the BSSO, we have analyzed the subject based on three main types of loading according to the tooth morphology. Molars are meant to bear heavy loading where as incisors are sharp ones used to cut and tear food and canines are at corners to cut food and position food in equilibrium. We also analyzed the airflow patterns in the upper pharyngeal airway at pre- and post-surgical conditions.

Baseline Model

There are three models which can be classified according to the loading condition (see Fig. 3). The mandible in the pre-surgical level is considered as the baseline model to compare with. Tooth displacements were interpreted in the vertical direction in accordance with loading direction.

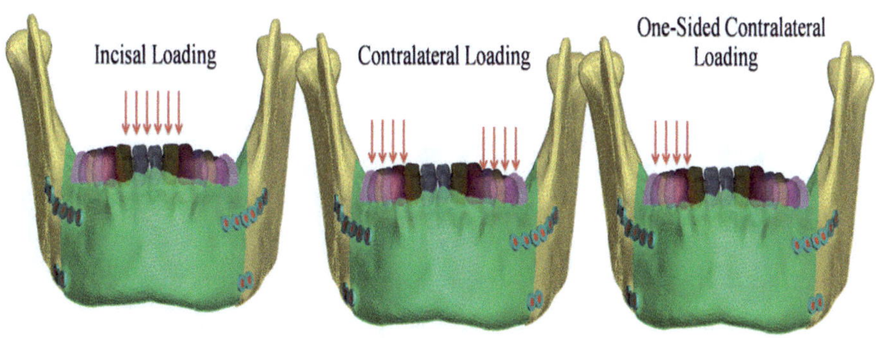

Fig. 3 Types of masticatory loads applied

5 mm BSSO Model

There are several factors that will determine the optimal modification for BSSO surgery like the position of the third molar, position if the mandibular foramen, and the alveolar nerve locations. We have modeled the BSSO model with all these surgical considerations as mentioned. The 5 mm advancement is one of the levels of mandibular advancements done in this study.

7 mm BSSO Model

The 7 mm advancement is the second level of advancement which is done in this study. We modeled perfect adaptation on the contacting surfaces with a viewpoint of high mechanical stability due to the complex biomechanical behavior of mandible and implant system.

Displacements Due to Loading, Maximum Bone Stress, Maximum Monocortical Miniplate Stress

Comparison of all the three loading conditions in the following baseline, 5 and 7 mm model is done in Figs. 3, 4 and 5 mentioned above. On comparing the three loading conditions maximum vertical deflection were found in the contralateral loading and one-sided contralateral loading with values nearly 1.9–2 mm, whereas incisal loading showed deflection of 1.7 mm (see Fig. 7). Miniplates and screws

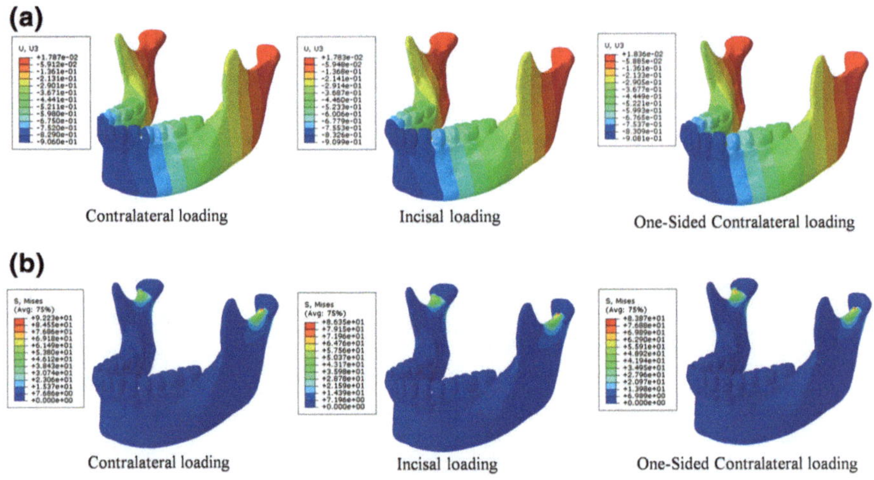

Fig. 4 **a** Vertical displacement and **b** bone stress (Von Mises stress)

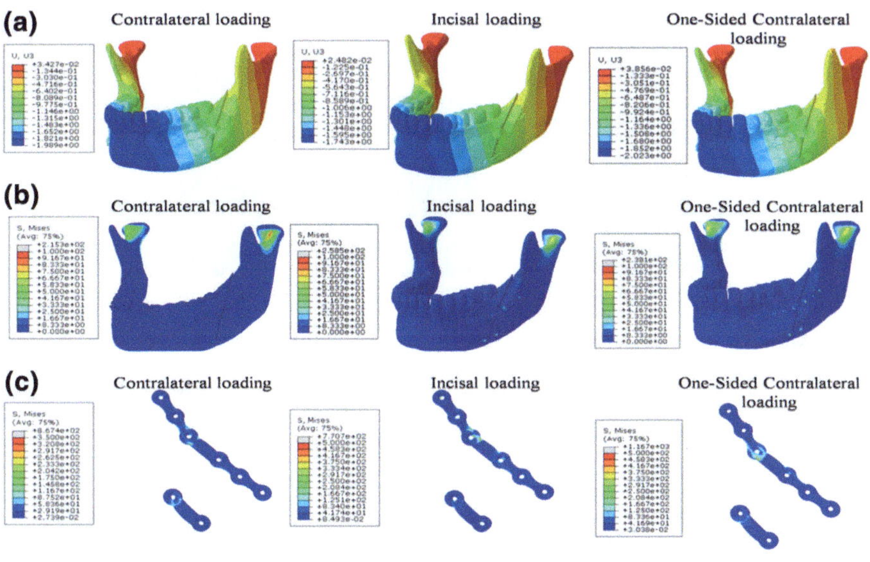

Fig. 5 **a** Vertical displacement, **b** stress at bone (Von Mises stress) and **c** stress at miniplates

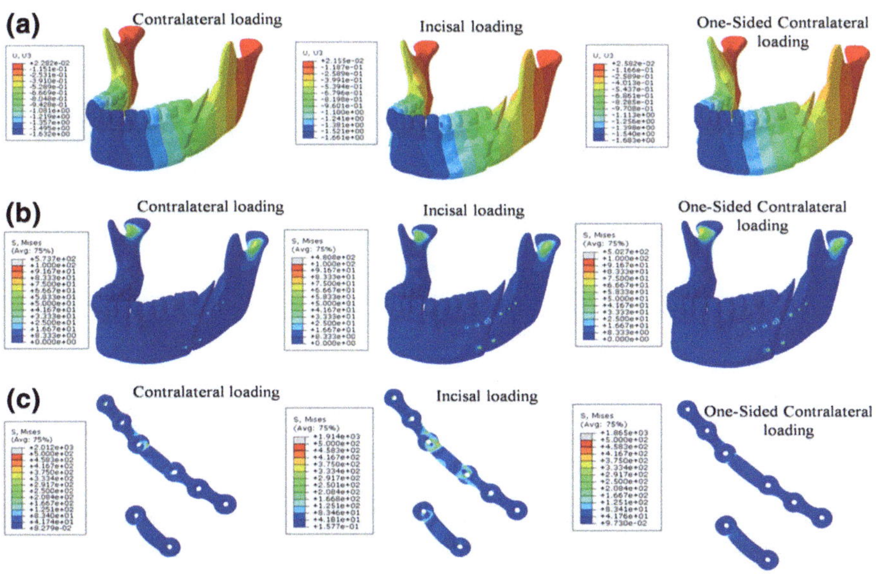

Fig. 6 **a** Vertical displacement, **b** stress at bone (Von Mises stress) and **c** stress at miniplates

were inferred with increase in stress with respect to increase in the level of advancement in BSSO (see Fig. 8). We can observe in Fig. 8 that maximum stress in miniplate is lower than material yield strength (895 MPa) and thus the implant had good strength to withstand the applied masticatory load (Fig. 6).

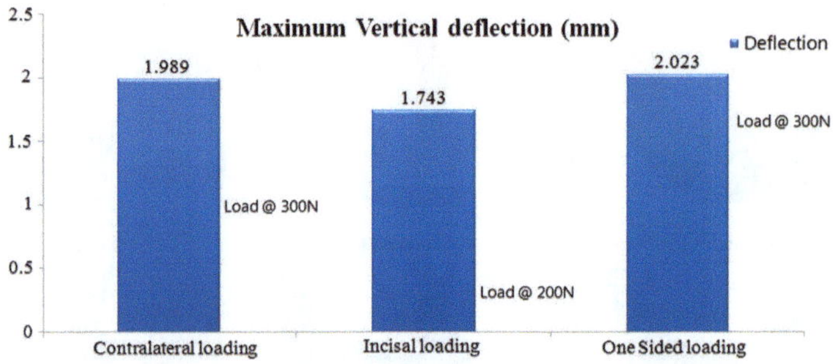

Fig. 7 Maximum vertical deflection

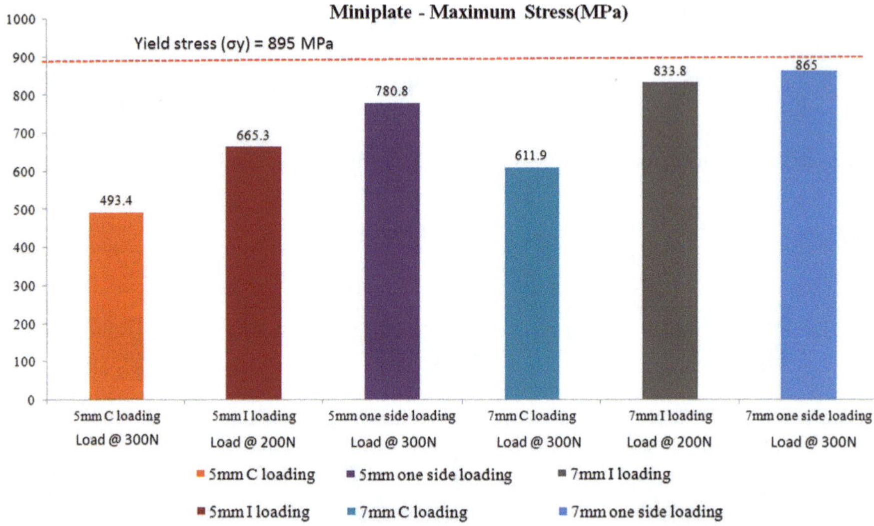

Fig. 8 Maximum miniplate stress

Airflow Dynamics in the Upper Airway Path

Considering the two different upper pharyngeal airway models, we can predict the pre- and post-operative airflow dynamics during inhalation. The extended airway model mimics the post-surgical airway path as the oropharyngeal region is expanded by considering the mandibular advancement. The velocity of airflow during inhalation increased in the areas of constriction in the upper airway volume near oropharynx. There were few areas of recirculation near the palatine tonsils and the laryngopharynx. Velocity reduces with the level of extension and increase of the cross-sectional area of the constriction. Pressure increased at the extended regions and few pressure drop areas occured due to recirculation zones (Figs. 9 and 10).

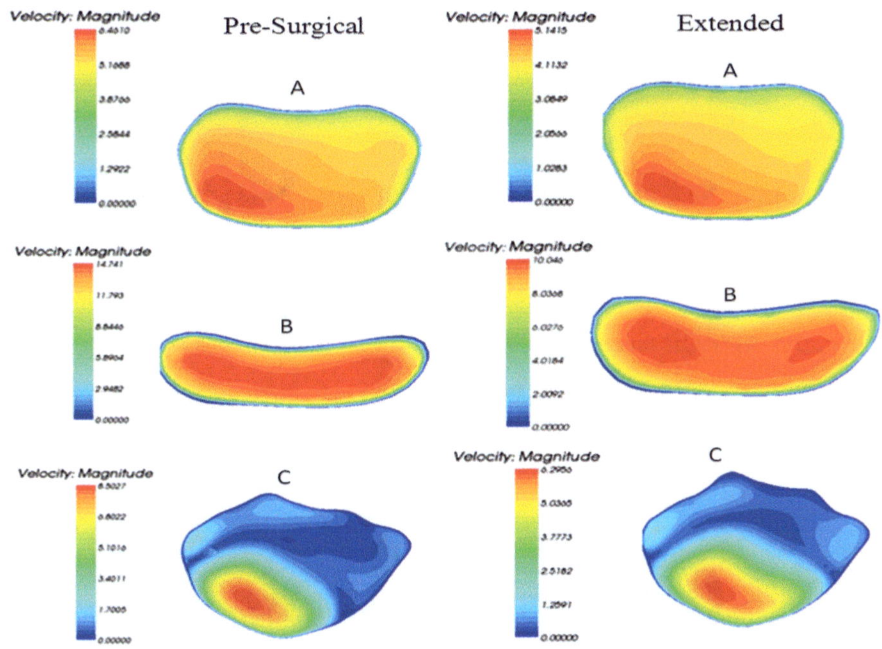

Fig. 9 Velocity contours on cross-sections

Discussion

The assessment of airway and its airflow dynamics is important to predict sleep disorders. Nowadays doctors and surgeons predict the breathing disorders only from the morphology of the airway path taken from the MRI sleep study or CT scan. The literature states that sleep disorders due to breathing are affected by factors like dental structures, bone structures, tongue volume, nasal morphology, and the airway volume. There is always lack of information on the flow patterns of air inside the upper airway path. The obstructive sleep apnea patients need refractive surgical management on jaw bone to improve the volume of air flowing into lower airway and its gas exchange.

Surgeons mostly perform cephalometric plan to enable orthodontic and maxillofacial surgical treatment's planning. But interestingly patients with normal cephalometric measurements sometimes had abnormal breathing rates and sleep disorders. Also it is clearly established that abnormal cephalometric measurements post-surgery are quite striking than the original pre-surgical cephalometric measurements.

The BSSO technique performed in our study is the most common orthognathic procedure for OSA patients. For modeling the BSSO surgery there are several factors that are maintained to attain significant placements of miniplates and screws. Basically osteotomy is decided according to the morphology of the bone and

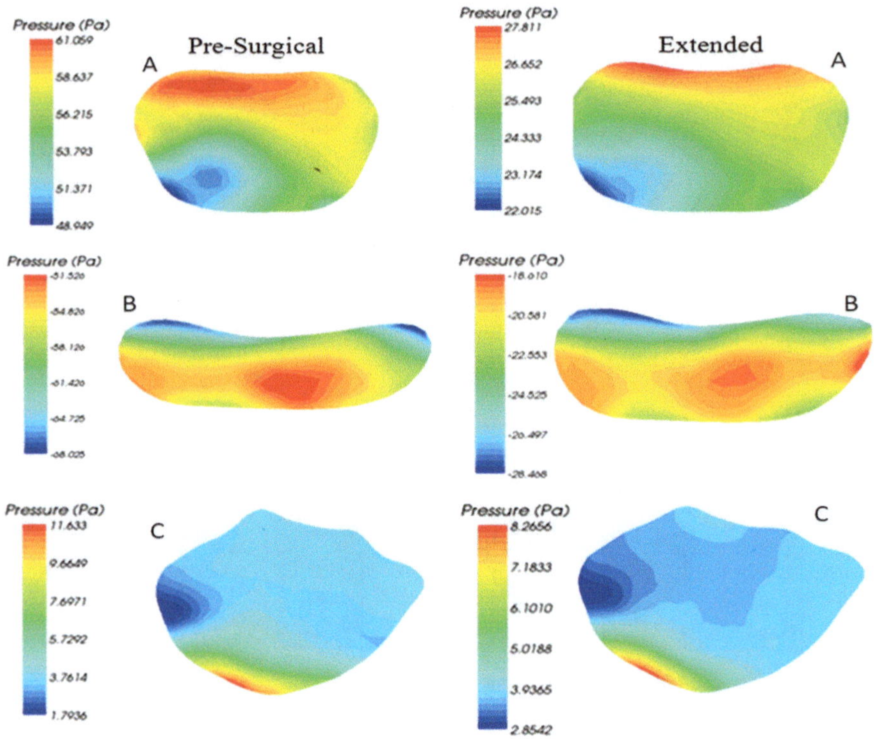

Fig. 10 Pressure contours on cross-sections

the need. In order to have better stability and reduced trauma surgeons has to focus on the location of the fixation system. We followed the Champy's lines which are the ideal line of osteosynthesis, performed to introduce stable fixation without risks due to screws and miniplates. Another advantage of Champy's lines is that we can find the exact area of placement without damaging the roots of the teeth. Volmer et al. have proved high correlation (>0.992) between FEA simulation and in vitro specimens under surgery.

We have shown the maximum vertical displacement in justice to the masticatory tooth loads. The idea of our study is to analyze a few major factors such as the exact fixation for the osteotomy, the level of masticatory loads which the implant system and the bone can withstand with less deformation and CFD of airflow patterns after the FEA in silico surgical setup. The expected result is said to have high stability in fixation as the displacement is less (<2.2 mm) when compared to literature and adding the fact that the implant system did not fail (see Fig. 8). A successful BSSO surgery for OSA patients is determined by the level of patency for airflow in airway after the surgery. In order to analyze the flow streamlines and the nature of flow after our virtual surgery for which we performed CFD of the upper airway region. The BSSO surgery helped to understand the airway patterns and the key areas of risk in which airway path gets constricted. When volume of airway path increases the level of breathing patency increases, thus leading to better sleep for OSA patients.

The most constricted areas of airway passages in the upper airway anatomy were the nasal cavity, nasopharynx, oropharynx, and hypopharynx. The model is extended in the areas of airflow challenges. We found that turbulence occurred at the nasal and oropharyngeal areas formed recirculation zones; as a result there was an increase in turbulent kinetic energy, velocity, and pressure at the constricted zone. At a flow rate of 15 l/min as given in this study is to understand the normal adult breathing patterns during rest. Due to the increase in the cross-sectional area in the plane C when compared to plane B the pressure drop is reduced and controlled to be equipoise between plane B and C. The plane B which is near the oropharyngeal region was found to be with higher ($\simeq$4.25 Pa) WSS due to largest frictional loss in flow. But this WSS is reduced for extended model ($\simeq$2.8 Pa). Therefore, we can agree to the fact that the increase in the advancement in BSSO can decrease the workload of breathing for OSA patients.

Conclusion

The stress and deflection at the implant–bone interface and the airflow dynamics before and after BSSO surgery was analyzed computationally. The actual surgical scenario was depicted and the implants were placed on the mandibular bone. The virtual BSSO surgery was performed in accordance with the standard surgical methods followed by maxillofacial surgeons. Masticatory loads were considered for assessing better fit and durability of the implant system. In addition to this, the airflow dynamics inside the upper pharyngeal airway volume was evaluated before and after mandibular advancement. Inhalation of air in the extended airway was found to have reasonable ranges of pressure, velocity, and wall shear stress as per the surgeon's expectation and better airflow results were inferred. Thus we suggest, virtual planning of BSSO paves way for prediction of possible airway patency that can be found at the pre-surgical level.

Acknowledgements We would like to thank for support and clinical guidance given to us. Dr. Manikandan BDS, MDS, FDSRCS, FFDRCSI, FTPRM Oral And MaxilloFacial Surgeon, Implantologist, Dentist, Meenakshi Ammal Dental College and Hospital.

References

Baek S-H, Cha H-S, Cha J-Y, Moon Y-S, Sung S-J (2012) Three-dimensional finite element analysis of the deformation of the human mandible: a preliminary study from the perspective of orthodontic mini-implant stability. Korean J Orthod 42(4):159–168

Biswas BK, Bag S, Pal S (2013) Biomechanical analysis of normal and implanted tooth using biting force measurement. ISSN 4:2305–8269

Champy M, Loddé JP, Schmitt R, Jaeger JH, Muster D (1978) Mandibular osteosynthesis by miniature screwed plates via a buccal approach. J Maxillofac Surg 6:14–21

Chun KJ, Choi HH, Lee JY (2014) Comparison of mechanical property and role between enamel and dentin in the human teeth. J Dent Biomech 5:1758736014520809

Erkmen E, Simsek B, Yücel E, Kurt A (2005) Three-dimensional finite element analysis used to compare methods of fixation after sagittal split ramus osteotomy: setback surgery-posterior loading. Br J Oral Maxillofac Surg 43:97–104

Feller BK-U, Schneider M, Hlawitschka M, Pfeifer G, Lauer G, Eckelt U (2003) Analysis of complications in fractures of the mandibular angle—a study with finite element computation and evaluation of data of 277 patients. J Craniomaxillofac Surg 31:290–295

Ghoneima A, AlBarakati S, Jiang F, Kula K, Wasfy T (2015) Computational fluid dynamics analysis of the upper airway after rapid maxillary expansion: a case report. Progr Orthod 16:10

Li KK, Riley RW, Powell NB, Guilleminault C (2000) Maxillomandibular advancement for persistent obstructive sleep apnea after phase-I surgery in patients without maxillomandibular deficiency. Laryngoscope 110:1684–1688

Phee BL (2015) Clinical outcomes of maxillo mandibular advancement surgery for the treatment of obstructive sleep apnea. Electronic thesis and dissertation repository, paper 2654

Sato FR, Asprino L, Consani S, Noritomi PY, de Moraes M (2012) A comparative evaluation of the hybrid technique for fixation of the sagittal split ramus osteotomy in mandibular advancement by mechanical, photoelastic, and finite element analysis. Oral Surg Oral Med Oral Pathol Oral Radiol 114(Suppl 5):S60–S68

Schendel SA, Broujerdi JA, Jacobson RL (2014) Three-dimensional upper-airway changes with maxillomandibular advancement for obstructive sleep apnea treatment. Am J Orthod Dentofacial Orthop 146:385–393

Stringhini DJ, Sommerfeld R, Uetanabaro LC, Leonardi DP, Araújo MR, Rebellato NL, da Costa DJ, Scariot R (2016) Resistance and stress finite element analysis of different types of fixation for mandibular orthognathic surgery. Braz Dent J 27(3):284–291

Takahashi H, Moriyama S, Furuta H, Matsunaga H, Sakamoto Y, Kikuta T (2010) Three lateral osteotomy designs for bilateral sagittal split osteotomy: biomechanical evaluation with three-dimensional finite element analysis. Head Face Med 6:4

Vollmer D, Meyer U, Joos U, Vègh A, Piffko J (2000) Experimental and finite element study of a human mandible. J Craniomaxillofac Surg 28:91–96

Wang Y, Elghobashi S (2014) On locating the obstruction in the upper airway via numerical simulation. Respir Physiol Neurobiol 193:1–10

Won CHJ, Li KK, Guilleminault C (2008) Surgical treatment of obstructive sleep apnea: upper airway and maxillomandibular surgery. Proc Am Thorac Soc 5:193–199

Wootton DM, Luo H, Persak SC, Sin S, McDonough JM, Isasi CR, Arens R (2014) Computational fluid dynamics endpoints to characterize obstructive sleep apnea syndrome in children. J Appl Physiol 116:104–112

A Novel Model for the Optimization of Load Carriage Ergonomics

Amir Hadid, Gal Gozes, Avihai Atoon, Amit Gefen and Yoram Epstein

Abstract Soldiers, recreational backpackers, and students are often required to carry their own equipment using backpacks. Shoulder strain is one of the limiting factors of load carriage, due to higher sensitivity for pressure hot spots, and susceptibility to short-term injuries such as soft tissue damage and trapped nerves or obstruction of blood vessels. However, to date, no optimized system is available for heavy load carriage, and the considerations for schoolchildren and students are more fashion-weighted than ergonomics or comfort-weighted.

Introduction

Soldiers, recreational backpackers, and students are often required to carry their own equipment using backpacks (Dean 2004; Knapik and Reynolds 2016; Negrini et al. 1999; Negrini 2002; Negrini et al. 2004; Sheir-Neiss et al. 2003). Shoulder strain is one of the limiting factors of load carriage, due to higher sensitivity for pressure hotspots, and susceptibility to short-term injuries such as soft tissue damage and trapped nerves or obstruction of blood vessels (Birrell and Hooper 2007; Wettenschwiler et al. 2015). However, to date, no optimized system is available for heavy load carriage, and the considerations for schoolchildren and students are more fashion-weighted than ergonomics or comfort-weighted (Fig. 1).

The aim of the current study was to develop a model for various loads and strap materials and structures in order to simulate real life loading scenarios and to help in optimizing load carriage systems design.

A. Hadid (✉) · G. Gozes · A. Atoon · A. Gefen
Department of Biomedical Engineering, Tel Aviv University,
6997801 Tel Aviv, Israel
e-mail: amir.hadid@gmail.com

Y. Epstein
Heller Institute of Medical Research, Sheba Medical Center, Ramat Gan, Israel

Y. Epstein
Sackler Faculty of Medicine, Tel Aviv University, 6997801 Tel Aviv, Israel

A. Gefen and D. Weihs (eds.), *Computer Methods in Biomechanics and Biomedical Engineering*, Lecture Notes in Bioengineering,
DOI 10.1007/978-3-319-59764-5_3

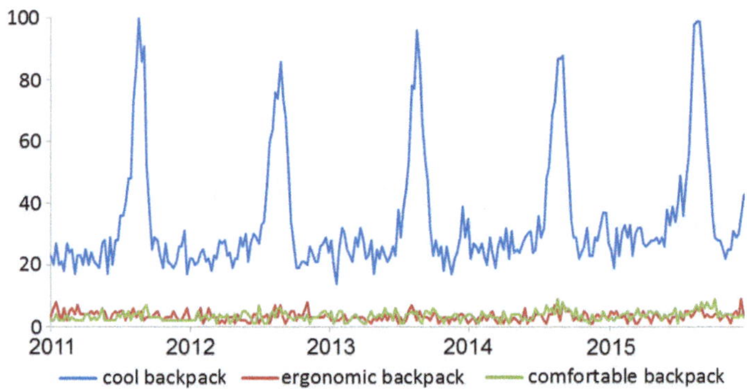

Fig. 1 Google Trends interest over time in 'cool backpack', compared to 'ergonomic backpack' and 'comfortable backpack'. *Numbers* represent search interest relative to the highest point on the chart for the given region and time. A value of 100 is the peak popularity for the term. A score of 0 means the term was less than 1% as popular as the peak. *Source* Google

Materials and Methods

Open-MRI scans were used for reconstructing a 3D geometrical model of an unloaded shoulder and for measuring the soft tissue deformations caused by a 25 kg backpack; subsequently, a subject-specific finite element (FE) stress–strain analyses was developed. Loads were applied at the strap–shoulder contact surface of the model by pulling the strap towards the shoulder until the desired load was attained. Strain at the subclavian artery, a marker for the brachial plexus region, served as the outcome measure for comparisons. The following material properties were evaluated: Homogeneous straight strap with varied material properties: $E = 0.5$, 1.2 and 5 MPa. Dual layer straight strap with $E = 20$ MPa for the inner layer and 0.5 MPa for the outer layer. Based on the results of the first simulations, a new anatomical strap was designed to maximize bone contact and avoid deformation on the soft tissues of the shoulder. It was compared to the straight strap (with the homogeneous single layer).

Results

A stiffer strap material ($E = 5$ MPa) resulted with a fourfold reduction in brachial plexus strains compared to the softer strap ($E = 0.5$ MPa). The dual strap structure resulted in brachial plexus strains, which were lower compared to the single layer $E = 0.5$ MPa strap. The anatomical strap design resulted in additional fourfold decrease in brachial plexus strains compared to the lowest strains obtained for the straight strap (Fig. 2).

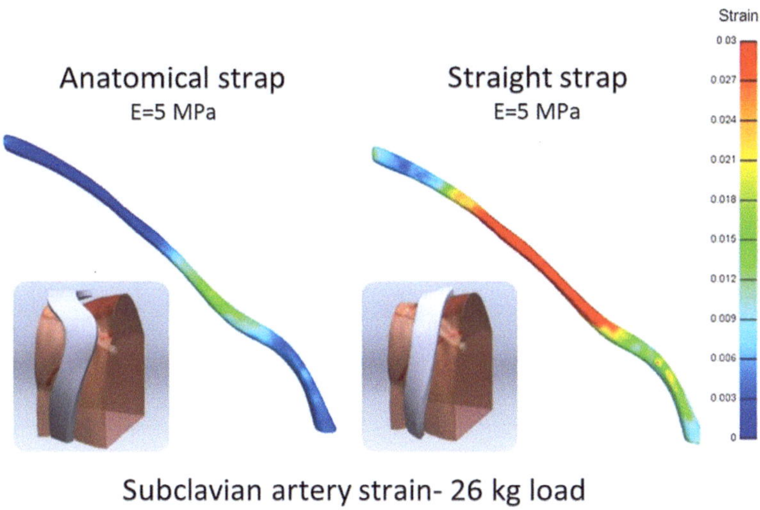

Fig. 2 Effective strains on the subclavian artery during 26 kg simulated load; *left* newly designed anatomical strap. *Right* straight strap

Discussion

The newly developed model successfully enabled predictions of soft tissue deformations in the subclavian artery and brachial plexus for different backpack strap materials and geometries. This model, verified by MRI measurements, can serve as a tool for load carriage optimization in order to reduce the straps-skin stresses and the consequent subclavian artery strains. Such optimization could improve the wearer comfort and the upper limb performance, by preventing the vascular and neurological impairments that are associated with load carriage, even for relatively light loads. The results from the current study indicate that an anatomical strap designed with appropriate materials can minimize the brachial plexus strains and the associated medical complications.

References

Birrell SA, Hooper RH (2007) Initial subjective load carriage injury data collected with interviews and questionnaires. Mil Med 172(3):306–311

Dean C (2004) The modern warrior's combat load. Dismounted operations in Afghanistan, April–May 2003. Army Center for Lessons Learned, Ft Leavenworth, Kan

Knapik JJ, Reynolds K (2016) Load carriage-related injury mechanisms, risk factors, and prevention. In: Gefen A, Epstein Y (eds) The mechanobiology and mechanophysiology of military-related injuries. Springer, Cham, pp 107–137

Negrini S (2002) Backpacks on! Schoolchildren's perceptions of load, associations with back pain and factors determining the load. Spine 27(2):187

Negrini S, Carabalona R, Sibilla P (1999) Backpack as a daily load for schoolchildren. Lancet 354(9194):1974

Negrini S et al (2004) The backpack load in schoolchildren: clinical and social importance, and efficacy of a community-based educational intervention. A prospective controlled cohort study. Eura Medicophys 40(3):185–190

Sheir-Neiss GI et al (2003) The association of backpack use and back pain in adolescents. Spine 28 (9):922–930

Wettenschwiler PD et al (2015) Mechanical predictors of discomfort during load carriage. PLoS ONE 10(11):e0142004

Breast Biomechanical Modeling for Compression Optimization in Digital Breast Tomosynthesis

Anna Mîra, Ann-Katherine Carton, Serge Muller and Yohan Payan

Abstract Mammography is a specific type of breast imaging that uses low-dose X-rays to detect cancer in early stage. During the exam, the women breast is compressed between two plates until a nearly uniform breast thickness is obtained. This technique improves image quality and reduces dose but can also be the source of discomfort and sometimes pain for the patient. Therefore, alternative techniques allowing reduced breast compression is of potential interest. The aim of this work is to develop a 3D biomechanical Finite Element (FE) breast model in order to analyze various breast compression strategies and their impact on image quality and radiation dose. Large breast deformations are simulated using this FE model with ANSYS software. A particular attention is granted to the computation of the residual stress in the model due to gravity and boundary conditions (thorax anatomy, position of the patient inside the MRI machine). Previously developed biomechanical breast models use a simplified breast anatomy by modeling adipose and fibroglandular tissues only (Rajagopal et al. in Wiley Interdiscip Rev: Syst Biol Med 2:293–304, 2010). However, breast reconstruction surgery has proven the importance of suspensory ligaments and breast fasciae on breast mechanics (Lockwood in Plast Reconstr Surg 103:1411–1420, 1999). We are therefore consider using a more realistic breast anatomy by including skin, muscles, and suspensory ligaments. The breast tissues are modeled as neo-Hookean materials. A physical correct modeling of the breast requires the knowledge of the stress-free breast configuration. Here, this undeformed shape (i.e., without any residual stress) is computed using the prediction–correction iterative scheme proposed by Eiben et al. (Ann of Biomed Eng 44:154–173, 2016). The unloading procedure uses the breast configuration in prone and supine position in order to find a unique displacement vector field induced by gravitational forces. The 3D breast geometry is reconstructed from MRI images that are segmented (Yushkevich et al. in Neuroimage 31:1116–1128, 2006) to differentiate the four main tissue types. The breast volume is discretized with a hexa-dominant FE

A. Mîra (✉) · Y. Payan
TIMC-IMAG Lab, CNRS, Université Grenoble Alpes, Grenoble, France
e-mail: anna.mira@imag.fr

A. Mîra · A.-K. Carton · S. Muller
GE-Healthcare, Buc, France

A. Gefen and D. Weihs (eds.), *Computer Methods in Biomechanics and Biomedical Engineering*, Lecture Notes in Bioengineering,
DOI 10.1007/978-3-319-59764-5_4

meshing tool as a unique volume. Finally, the model is evaluated by comparing the estimated breast deformations under gravity load with the experimental ones measured in three body positions: prone, supine, and oblique supine.

Introduction

Today, projection mammography is the key imaging modality for breast cancer screening and plays an important role in diagnostics. During the exam, the women breast is compressed between two plates until a nearly uniform breast thickness is obtained. Breast compression improves image quality and reduces the absorbed dose of ionizing photons. But breast compression can also be the source of discomfort and sometimes painful for the patient during and after the exam. The discomfort perceived during the exam could deter women from getting the exam. Therefore, an alternative technique with reduced breast compression is of potential interest.

The aim of this work is to develop a biomechanical finite element (FE) breast model allowing the investigation of alternative breast compression strategies. Ultimately, their impact on mammography image quality and radiation dose could be investigated using a simulation framework generating numerical breast-like phantoms, FE mammographic breast deformation, and image acquisitions. The 3D strain/stress cartography derived from the FE solution could be used as a first measure of pain and discomfort.

Previously developed biomechanical breast models have used a simplified breast anatomy by modeling adipose and fibroglandular tissues only (ANSYS®). However, breast reconstruction surgery has proven the importance of suspensory ligaments and breast fasciae on breast mechanics (Hipwell et al. 2016). We are considering using a more complex breast anatomy by including breast ligaments as well as the skin and muscles.

A physical correct modeling of the breast deformation requires the knowledge of the stress-free breast configuration; i.e., the absence of gravity load. Here, the stress-free shape was computed using the prediction–correction iterative scheme (Carter et al. 2012), the unloading procedure used the breast configuration in prone and supine position to find a unique displacement vector field induced by gravitational forces. Finally, the estimated breast deformations under gravity load were compared with the experimental ones obtained from the MR images.

Materials and Methods

Data Acquisition and Preprocessing

The biomechanical breast model is built using MR images of two women between 50 and 55 years old and with respectively small and large breasts. Both women participated on a voluntary basis and signed an informed consent form approved by

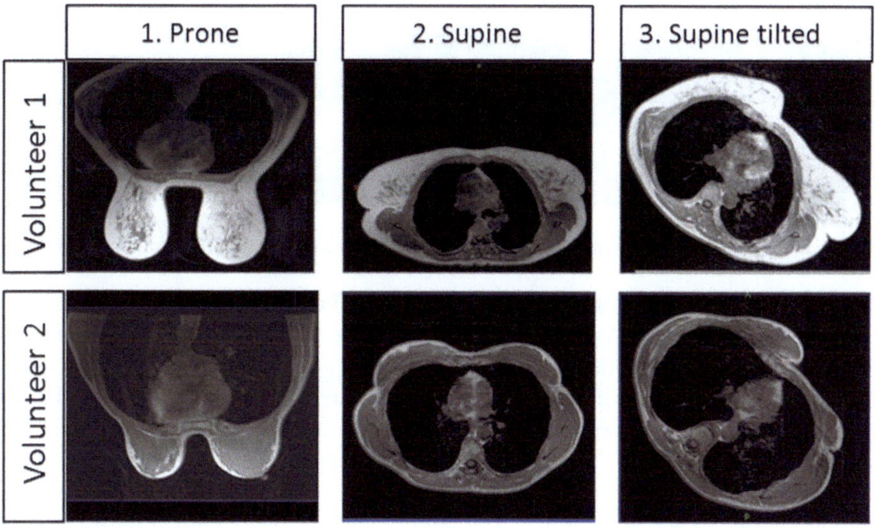

Fig. 1 Breast configurations under gravity loads

the ethics committee. The MRI was performed at 3T (Achieva 3.0TX, Philips, NL, USA) at IRMaGe MRI facility (Grenoble, France). Both women were imaged in three configurations: prone, supine, and supine tilted positions (Fig. 1).

The positions were chosen to assess the largest possible breast deformation without any contact area between the patient and the MRI scanner bore. Before image acquisition, ten fiducial markers were fixed on the skin surface; four on the chest wall and six on the breast skin. The markers were used to aid preprocessing registration of the breast volume and to validate the modeled breast deformation.

Image preprocessing starts with an image registration step. The rigid transformation between two breast configurations was computed using the marker's coordinates and the chest wall position in the three images. We assumed that only the four chest-wall markers affect the rigid transformation between breast configurations. Next, the MRI data were classified according to three tissue types: muscle, breast, and skin (Fig. 2). The segmented data set was then used to reconstruct the three-dimensional breast geometry. The image preprocessing steps were performed using ITK libraries VERSION_4.3.2 and ITK-Snap software VERSION_3.4.0 (Yushkevich et al. 2006).

Patient-Specific Finite Element Model (PSFEM)

From the segmented data, a tetrahedral finite element mesh was generated using ANSYS meshing tool (ANSYS®). Two components were created: one representing the breast soft tissues and its surroundings and the other representing the pectoral muscle (Fig. 2). The contact surface between the two components was defined as a

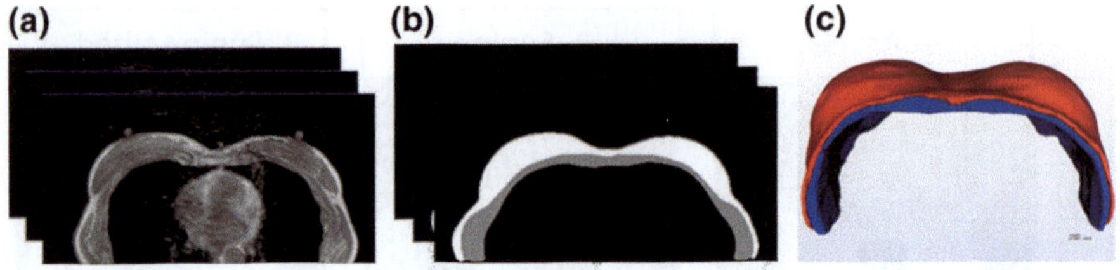

Fig. 2 **a** MRI data for supine position, **b** segmented data and **c** corresponding finite element mesh

"no-separation" contact (the two components are bonded) with an infinite friction coefficient (ANSYS®). As the skin is a thin layer organ (thickness of 2 mm) we used triangular shell elements to represent it.

Finally, the new finite element mesh was subjected to a hyper-elastic quasi-static simulation. Soft tissues were modeled as quasi-incompressible (Poisson ratio = 0.45) neo-Hookean materials. The optimal elastic parameters were determined by minimizing the error between the simulated and the experimental data, using a manual dichotomy on the elastic modulus. The search intervals were defined for each tissue type based on the data found in the previous studies (Hipwell et al. 2016; Krouskop et al. 1998; Rajagopal et al. 2008). The elastic parameter of the pectoral muscle was set to 30 kPa and the elastic parameters of breast tissues and skin were varied from 0 to 30 kPa and from 5 to 80 kPa respectively.

Stress-Free Geometry Estimation

In this work, the stress-free geometry was computed using a prediction–correction iterative scheme. The algorithm first proposed by Carter et al. (2012) was improved and adapted to our problem.

First, we assumed zero internal stress for the breast supine configuration generated from MR images (Fig. 3a Geometry of measured supine configuration). Then, the gravity load was applied in the reverse direction to give a first estimation of initial stress-free state (Fig. 3b Geometry of estimated stress-free configuration). Next, the gravity load was applied to the stress-free geometry to estimate the breast geometry in prone configuration (Fig. 3c). At this step, we defined a group of nodes (active nodes) which were used to compute the difference between the nodes position in estimated and measured prone configuration. The nodes were selected at the external surface of the breast. This active node difference was then applied to the stress-free geometry. The process was repeated until convergence was achieved.

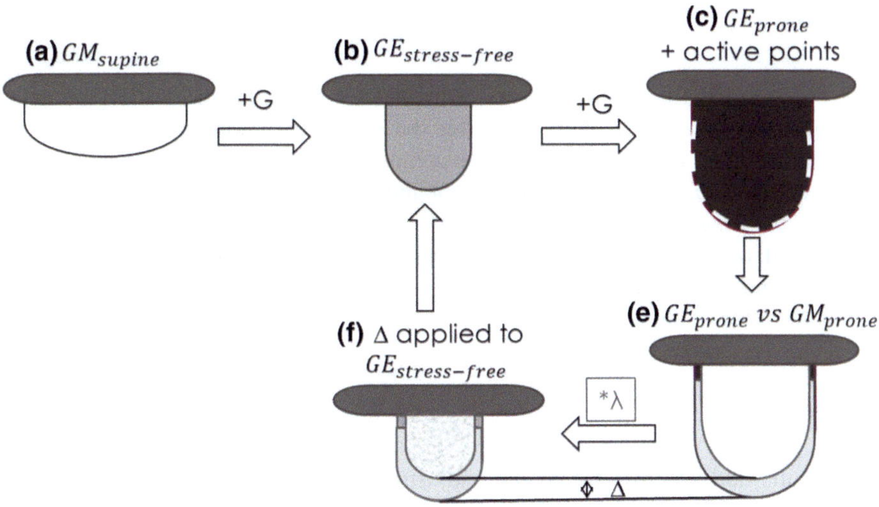

Fig. 3 Prediction–correction iterative scheme for stress-free geometry estimation. *GM* measured breast geometry; *GE* estimated breast geometry, Δ difference between estimated and measured geometries; λ regularization factor

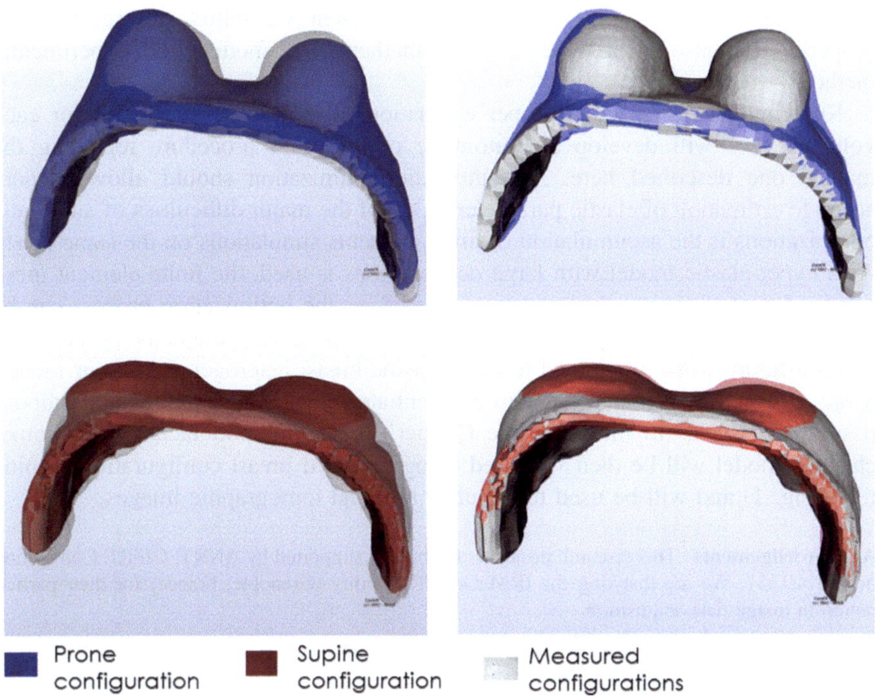

Fig. 4 Comparison of simulated configurations with the measured ones

Results

The breast stress-free geometry was estimated for each pair of elastic parameters defined from the search intervals. After the stress-free geometry estimation, the prone and supine breast positions were computed to evaluate the model. The best fitting of real data in both positions is obtained using an elastic parameter equal to 0.3 kPa for breast tissue and equal to 10 kPa for skin (see Fig. 4). The manual adjustment of elastic parameters conducted to breast tissues is softer than most values found in the literature.

Discussion

The results derived from the manual adjustment of tissues elasticity revealed that our breast model is highly sensitive on the value of elastic parameters. Therefore, a good estimation of the breast mechanics requires a precise estimation of tissues elasticity for each individual volunteer. We should note, however, that our results are only consistent with models developed and validated with data using MR images. Most of published papers on breast tissue elasticity propose very high values for fat, glandular and skin elastic parameters (Rajagopal et al. 2008). The large difference may come from the inter-women variability of breast elastic properties but also by a wide range of mathematical models and experimental methods.

Knowing that we need a proper estimation of the elastic properties for each volunteer, we will develop an automatic optimization procedure replacing the manual one described here. An automatic optimization should allow a more accurate estimation of elastic parameters. One of the major difficulties of automatic optimizations is the accumulation of finite elements simulations on the same mesh. As a hyper-elastic model with large deformations is used, the finite element mesh can be distorted after each simulation. Therefore, the optimization process can be stopped due to the poor-quality elements without achieving the convergence.

As a future work, we intend to consider the breast heterogeneity in our model. More specifically, it is necessary to differentiate the glandular tissues and adipose tissues, as well as to introduce the Cooper's ligaments and facias. Our biomechanical model will be then validated using the third breast configuration (supine tilted Fig. 1) and will be used to simulate artificial tomographic images.

Acknowledgements This research project is financially supported by ANRT, CIFRE Convention no. 2014/1357. We are thanking the IRMaGe MRI facility (Grenoble, France) for their participation in image data acquisition.

References

Bolt Automatic Hex Meshing, Computational Simulation Software. http://www.csimsoft.com/boltoverview.jsp

Carter T, et al (2012) Application of biomechanical modelling to image-guided breast surgery. In: Soft tissue biomechanical modeling for computer assisted surgery. Springer, Berlin, pp 71–94

Eiben B, Vavourakis V, Hipwell JH, Kabus S, Buelow T, Lorenz C, Hawkes DJ (2016) Symmetric biomechanically guided prone-to-supine breast image registration. Ann of Biomed Eng 44 (1):154–173

Hipwell JH et al (2016) A review of biomechanically informed breast image registration. Phys Med Biol 61(2):R1

Krouskop TA et al (1998) Elastic moduli of breast and prostate tissues under compression. Ultrason Imaging 20(4):260–274

Lockwood T (1999) Reduction mammaplasty and mastopexy with superficial fascial system suspension. Plast Reconstr Surg 103(5):1411–1420

Rajagopal V et al (2008) Creating individual-specific biomechanical models of the breast for medical image analysis. Acad Radiol 15(11):1425–1436

Rajagopal V, Nielsen PM, Nash MP (2010) Modeling breast biomechanics for multi-modal image analysis—successes and challenges. Wiley Interdiscip Rev: Syst Biol Med 2(3):293–304

ANSYS® Academic Research, Release 16.2, Help System, ANSYS Mechanical APDL Contact Technology Guide, ANSYS, Inc

ANSYS® Academic Research, Release 16.2, Help System, ANSYS Meshing, ANSYS, Inc

Yushkevich PA, Piven J, Hazlett HC, Smith RG, Ho S, James Gee C, Gerig G (2006) User-guided 3D active contour segmentation of anatomical structures: significantly improved efficiency and reliability. Neuroimage 31(3):1116–1128

Device-Related Pressure Ulcers from a Biomechanical Perspective

Ayelet Haimy, Kara Kopplin and Amit Gefen

Abstract A pressure ulcer (PU) is a localized injury of weight-bearing soft tissues, typically over a bony prominence, which develops primarily due to pressure, or pressure in combination with shear.

Introduction

A pressure ulcer (PU) is a localized injury of weight-bearing soft tissues, typically over a bony prominence, which develops primarily due to pressure, or pressure in combination with shear (Gefen 2007; National Pressure Ulcer Advisory Panel 2014). The elderly, frail, and individuals with impaired mobility and sensory capacities are well known to be prone to PUs; however, PUs in the pediatric population are inherently different from those in adults, in their risk factors and etiology. In the pediatric population there are unique mobility and spasticity issues, such as cerebral palsy or other congenital central nervous system dysfunctions, which introduce inherently different risk factors compared to adults (Coleman et al. 2013; Butler 2006; Levy et al. 2015; Baharestani and Ratliff 2007). Importantly, and distinguishably from adults, more than 50% of the PUs in children and neonates are related to the use or contact with medical equipment at the care setting (Baharestani and Ratliff 2007; Willock et al. 2005). While recent advances in acute care (e.g., in heart–lung support and neurosurgery) significantly increase the survival rates of preterm neonates, infants and children, they also expose them to the high-risk neonatal and pediatric care environments (Gefen 2015). When a stiff, curved object, such as an electrode, tube, catheter or wire, is wedged between the skin of an infant or child and the support surface (and especially if that happens near

A. Haimy · A. Gefen (✉)
Department of Biomedical Engineering, Faculty of Engineering,
Tel Aviv University, Tel Aviv, Israel
e-mail: ayeletlevy1@gmail.com

K. Kopplin
Efficacy and Research, ROHO, Inc, Belleville, IL, USA

A. Gefen and D. Weihs (eds.), *Computer Methods in Biomechanics and Biomedical Engineering*, Lecture Notes in Bioengineering,
DOI 10.1007/978-3-319-59764-5_5

a weight-bearing bony prominence), elevated localized deformations occur in the skin and underlying soft tissues. The sustained exposure to such deformations may jeopardize the integrity and viability of these highly distorted tissues, and initiate a deformation-induced injury, which is later exacerbated by localized ischemia and blockage of the lymphatics (Levy et al. 2015). The aims of this work were (i) to determine the mechanical states in the scalp of a newborn lying supine, near a wedged EEG electrode or wire, which is deforming the scalp at the occiput, and (ii) to evaluate the effect of a doughnut-shaped headrest on the mechanical state of tissues at the same site.

Materials and Methods

We employed finite element (FE) computational modeling to evaluate the mechanical states of skin and fat tissues at the occipital region of a newborn when the scalp interacts with a medical device wedged between the skin and the mattress. First, we used a portable three-dimensional laser scanner (Sense™) to acquire the outer shape and dimensions of a Baby-born® doll head. Then, we used the +CAD module of Simpleware® software suite (2012) to convert the.stl file generated by the laser scanner to a voxel-based array database and then loaded the robust geometrical model of the head into the ScanIP module of the Simpleware® software suite. Then, we performed segmentation and meshing of the skin, fat and skull in ScanIP of Simpleware®. Next, we added the geometries of a standard foam mattress, a disk-shaped electrode, a cylindrical wire and a doughnut-shaped headrest in Preview module of FEBio. Constitutive laws and mechanical properties of all the model components were adopted from the literature. Specifically, the skull, flat foam mattress, doughnut-shaped headrest, electrode, and wire were assumed to be isotropic linear-elastic materials while fat and skin tissues were assumed to be nearly incompressible hyperelastic isotropic materials with their large deformation behavior described using an uncoupled neo-Hookean material model (FEBio 2012). Boundary conditions were then chosen to simulate weight bearing of the infant's head during supine lying.

For validation of our computational modeling, we have reconstructed the doughnut-shaped headrest model variant in the physical world, using a phantom of a baby's head embedded with sensors (Fig. 1). First, we used dry rice to fill the head of the Baby-born doll (same head that has been scanned for the modeling) at high precision, until reaching a weight of 400 g. Then, we casted paraffin gel in a doughnut-backing tray to mold the doughnut-shaped headrest (mean diameter 60 mm, tube diameter 30 mm). We used four piezoelectric force sensors (FlexiForce® A401 by Tekscan Inc., Boston, MA, USA) positioned at equal spaces along the perimeter of the doughnut to measure the weight force transferred through the four quadrants of the headrest, with the doll head in place (Fig. 1). The average weight force calculated from the two medial quadrants and the two lateral quadrants of the FE model were 125.7 and 67.4 g, respectively, and the corresponding values measured from the sensors in the phantom system were 122.7 and 71.4 g, which is

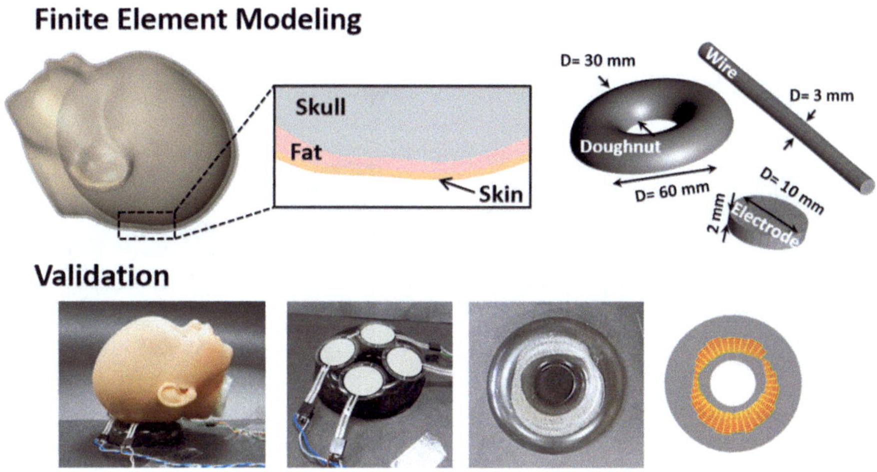

Fig. 1 The three-dimensional anatomical model of the skull, fat and skin tissues and the modeled intrusive medical devices that were tested (*top*). Validation of the modeling using a phantom of a baby's head on a jelly doughnut embedded with force sensors (*bottom*)

in excellent agreement. Then, using water-based paint applied on the back of the doll's head, we recorded an imprint of the actual contact area on the doughnut, and have digitized the image of the stained doughnut. Differences between the contact area calculated from the FE model and that measured in the phantom system were <2%, which again indicates excellent agreement and which, together with the force measurements, validates the modeling (Fig. 2b). We examined contact pressures, effective (von Mises) stresses, shear stresses and strain energy density (SED) in the fat and skin tissues at the occipital region.

Results

While occipital pressures in the reference model configuration reached a maximal value of 8.3 kPa, corresponding peak values for the case of interface with the electrode, wire, and doughnut-shaped headrest reached 13.7, 114.8, and 12.9 kPa, respectively. Additionally, we found a 4.5%-increase in the total contact area for the case where a doughnut-shaped headrest was used with respect to the reference flat support case, but as said above, despite that slightly increased contact area, the peak pressure actually increased for the doughnut case (Fig. 2). SED distributions in the soft tissue around the occiput are shown in Fig. 2. An increased SED was found in the deep fat tissue near the interface with the occipital bone, consistently in all of the model variants. Overall, the lodged wire resulted in the worse mechanical conditions in the soft tissues, compared to the lodged electrode and use of a doughnut-shaped headrest, with 345 and 50% increase in effective stresses in skin and fat tissues, respectively. The wedged electrode resulted in 60 and 368%

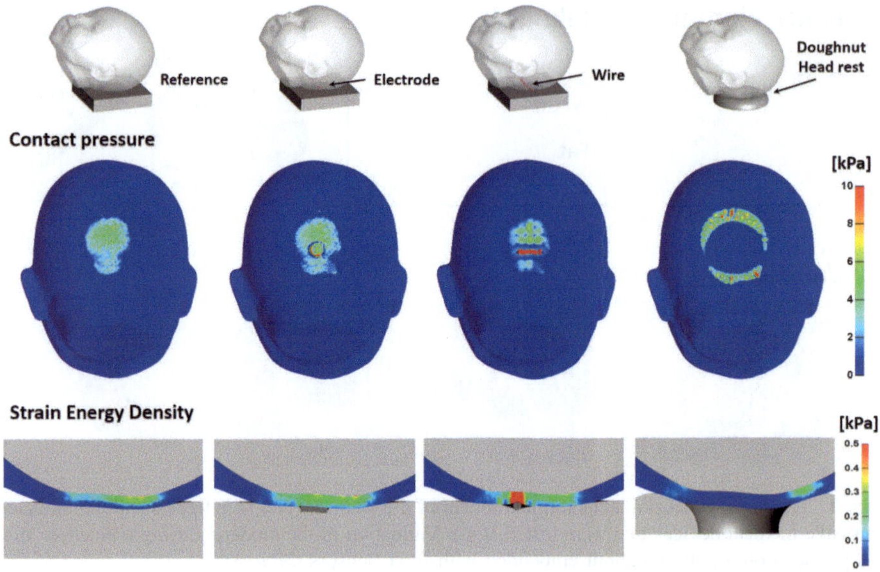

Fig. 2 Comparison of contact pressures on the skin of the occipital region, during supine weight bearing, on a flat foam mattress, with an electrode wedged between the skin and the mattress, with a wire wedged between the skin and the mattress and on a doughnut-shaped headrest (*top*). Comparison of strain energy densities in fat and skin tissues around the occiput during supine weight bearing, on a flat foam mattress, with an electrode wedged between the skin and the mattress, with a wire wedged between the skin and the mattress and on a doughnut-shaped headrest (*bottom*)

increase in peak SED values in the fat and skin tissues, respectively. Positioning the head of the baby on the doughnut-shaped headrest had a mild effect on the SED in fat tissues. However, it resulted in a 547% increase in SED in skin tissue. The interfering wire clearly inflicted the most substantial increase of SED in both skin and fat tissues, with 5.28- and 78-fold increase of peak SED values in the fat and skin tissues, respectively (Fig. 2).

Discussion

Overall, the wedged electrode and wire resulted in concentrated stresses and elevated SED, in the affected skin and fat tissues, with a substantially more prominent effect in the case of the wedged wire. This is likely due to the small diameter and high curvature (which can essentially be described as sharpness) of these thin wires that are compressed against the skin and locally distort it when bodyweight applies. Hence, in light of our present findings, we surmise that the geometry of the device has a critical effect on the level of mechanical interference with soft tissues in cases of misuse or misplacement. Interestingly, our present modeling study indicates that a doughnut-shaped headrest may not always improve the mechanical states in the affected tissues, since that simulation resulted in increased SED and stress values, but the topic needs further investigation incorporating additional headrest design variants. Considering that elevated and localized tissue deformations, stresses and SED indicate a risk for PUs, our simulations suggest that misplaced medical devices, and using a doughnut-shaped headrest, impose an actual risk for developing device-related PUs. Guidelines for pediatric clinical care should recommend routine inspection of the medical device placement to prevent harmful contact conditions with the patient. Furthermore, the improved design of medical equipment for pediatric settings is needed in order to protect these fragile young patients from PUs.

References

Baharestani MM, Ratliff CR (2007) Pressure ulcers in neonates and children: an NPUAP white paper. Adv Skin Wound Care 20(4):208–220

Butler CT (2006) Pediatric skin care: guidelines for assessment, prevention, and treatment. Pediatr Nurs 32(5):443–450

Coleman S, Gorecki C, Nelson EA, Close SJ, Defloor T, Halfens R, Farrin A, Brown J, Schoonhoven L, Nixon J (2013) Patient risk factors for pressure ulcer development: systematic review. Int J Nurs Stud 50:974–1003

FEBio (2012) Finite element for biomechanics, Theory manual ver. 1.5

Gefen A (2007) The biomechanics of sitting-acquired pressure ulcers in patients with spinal cord injury or lesions. Int Wound J 4(3):222–231

Gefen A (2015) Biomechanics of PUs in paediatric care settings. J Wound Care 24(3):81

Levy A, Kopplin K, Gefen A (2015) Adjustability and adaptability are critical characteristics of pediatric support surfaces. Adv Wound Care 4(10):615–622

National Pressure Ulcer Advisory Panel (2014) European pressure ulcer advisory panel and pan pacific pressure injury alliance. In: Haesler E (ed) Prevention and treatment of pressure ulcers: quick reference guide. Cambridge Media, Osborne Park

Simpleware® Ltd. (2012) ScanIP, +FE, +NURBS and +CAD Reference Guide ver. 5.1, 2012. http://www.simpleware.com/software/

Willock J, Harris C, Harrison J, Poole C (2005) Identifying the characteristics of children with pressure ulcers. Nurs Times 101(11):40–43

Comparison of Anisotropic Models to Simulate the Mechanical Response of Facial Skin

Cormac Flynn, Andrew T. Taberner, Sidney Fels and Poul M.F. Nielsen

Abstract Physically-realistic models of the face can contribute to development in several fields including biomedicine, computer animation, and forensics. Face models have benefited from better anatomical representation of the mimetic muscles, and more realistic interactions between soft and bony tissues. These models can also benefit from improved characterisation of the skin layer by having more authentic deformation and wrinkling behaviour. The objective of this work is to compare and evaluate the ability of different constitutive models to simulate the mechanical response of facial skin subjected to a rich set of deformations using a probe. We developed a finite element model to simulate facial skin experiments. Several anisotropic constitutive equations were tested for their suitability to represent facial skin. The finite element model simulated the force-displacement response of facial skin under a rich set of deformations. The variance accounted for between the experimental data and model data ranged from 79% for the Gasser et al. (2006) model to 96% for the Bischoff et al. (2002) model. Estimated pre-stresses ranged from 7 kPa in the lip region to 53 kPa in the central cheek region.

Introduction

Physically realistic biomechanical models of the face can be applied in a wide range of domains, including biomedicine, computer animation and forensics. There has been a continuous improvement in the anatomical accuracy of face models with

C. Flynn (✉)
Centre for Engineering and Industrial Design, Wintec, Hamilton, New Zealand
e-mail: cormac.flynn@wintec.ac.nz

A.T. Taberner · P.M.F. Nielsen
Department of Engineering Science, Auckland Bioengineering Institute, The University of Auckland, Auckland, New Zealand

S. Fels
Department of Electrical and Computer Engineering, University of British Columbia, Vancouver, Canada

A. Gefen and D. Weihs (eds.), *Computer Methods in Biomechanics and Biomedical Engineering*, Lecture Notes in Bioengineering,
DOI 10.1007/978-3-319-59764-5_6

better representation of the mimetic muscles, and realistic contact and attachments between soft and bony tissues (Flynn et al. 2015; Wu et al. 2014). Face simulations can also benefit from improved constitutive models of the skin layer. For example, better representation of the mechanical properties of facial skin can lead to improved predictions of deformations as a result of maxillofacial surgical procedures (Zhang et al. 2016).

There is a relative scarcity of facial skin model parameters in the literature. Characterisation studies using suction measurements include Weickenmeier et al. (2015), Luboz et al. (2014), and Ohshima et al. (2011). These approaches do not characterise the anisotropic properties of facial tissues due to the axi-symmetrical nature of the experimental protocol. Ohshima et al. (2011) also used a Reviscometer that characterised the anisotropy of facial skin but this approach assumed skin to have a linear stress–strain response. Few studies attempt to estimate the in vivo pre-stress in facial skin (Flynn et al. 2013). This pre-stress has a significant influence on the behaviour of skin and its inclusion would be an important development in improving the realism of any face model (Swain and Gupta 2015).

This work is a development of the model presented in Flynn et al. (2013). Specifically, the objective is to compare and evaluate several constitutive models and their ability to simulate the mechanical response of facial skin undergoing a rich set of deformations. Material parameter sets will be determined for these constitutive models in addition to estimates of the in vivo pre-stress that is present in facial skin.

Materials and Methods

In Vivo Facial Experiments

The facial skin of several volunteers was subjected to a multi-directional deformation set using a force-sensitive micro-robot (Fig. 1). The experiments are detailed in Flynn et al. (2013). To summarise, we attached a boundary ring to the volunteer's face, centring it on the point of interest. The volunteer rested his head on a support plate (Fig. 2). A 4 mm diameter cylindrical probe was positioned underneath the support plate. A key on the boundary ring matched a hole on the support plate, thus ensuring the orientation of the face with respect to the probe on the micro-robot was known. The probe was attached to the surface of the skin using liquid cyanoacrylate adhesive. We tested several points on the face (Fig. 3). For each location, the probe was moved in 16 different in-plane and out-of-plane directions (Fig. 4). For each direction, three triangular wave cycles of frequency 0.1 Hz were performed to pre-condition the skin. The time, probe displacement, and probe reaction force were recorded via a LabVIEW software interface (National Instruments, Austin, USA).

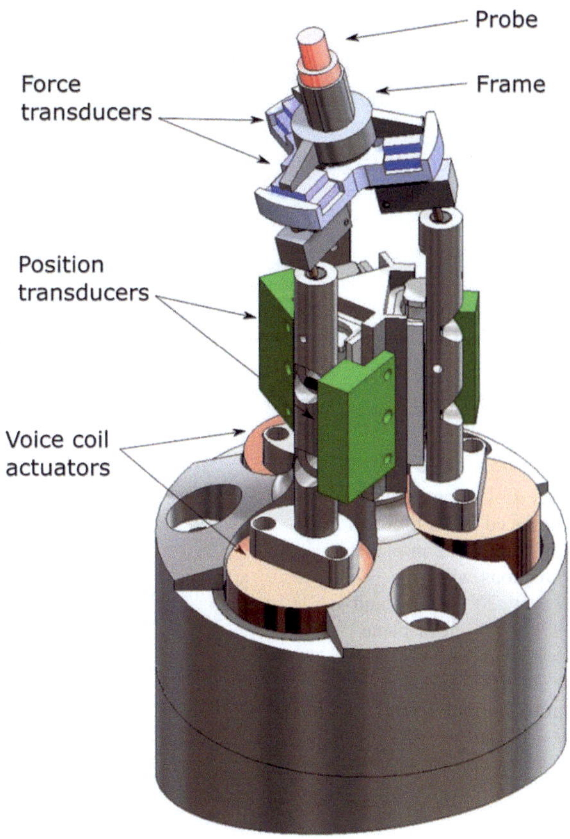

Fig. 1 Force-sensitive micro-robot

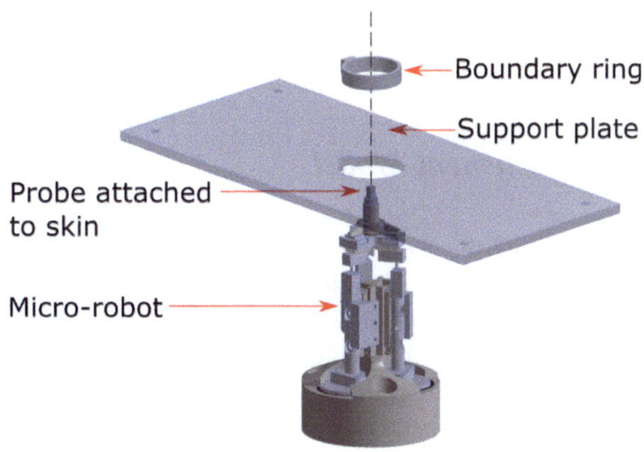

Fig. 2 Experimental setup: the boundary ring is attached to the volunteer's face and centred at one of the locations indicated in Fig. 3

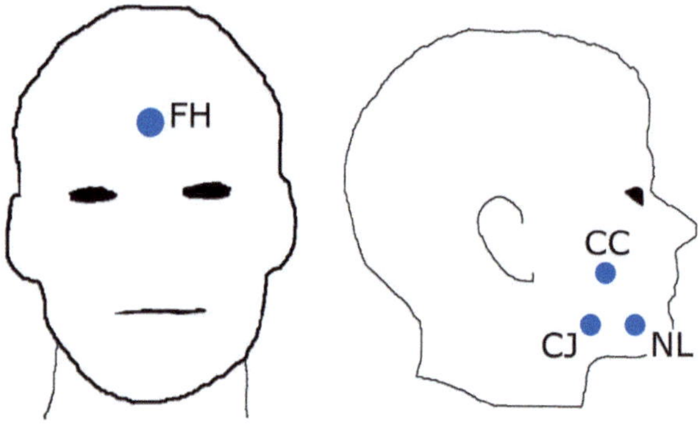

Fig. 3 Locations where the micro-robot probe was attached. *FH* forehead, *CJ* central jaw, *CC* central cheek, *NL* near lip

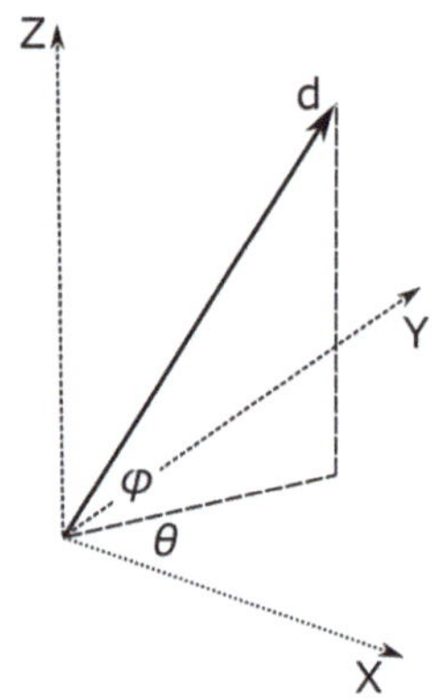

Fig. 4 The probe was displaced in 16 different directions—12 in-plane directions with $\theta = 0°, 30° \ldots 330°$ and $\phi = 0°$, 3 out-of-plane directions with $\theta = 0°, 45°, 90°$ and $\phi = 45°$ and 1 normal direction with $\theta = 0°$ and $\phi = 90°$. $d \approx 1.4$ mm for all directions

Finite Element Model

To simulate the facial skin experiments, we developed a finite element model in FEBio, a package specifically developed to perform nonlinear large deformation analyses in biomechanics (Maas et al. 2012). We modelled the skin as a square domain of side 50 mm (Fig. 5). Two circular partitions of diameter 37.5 and 4 mm were created representing the inside edge of the boundary ring and the outside edge of the probe, respectively. We created the mesh using Gmsh, a three-dimensional finite element mesh generator (Geuzaine and Remacle 2009). It consisted of 2432 quadrilateral shell elements. The thickness of the shell elements was 1.5 mm. A study was undertaken to ensure the mesh density was adequate.

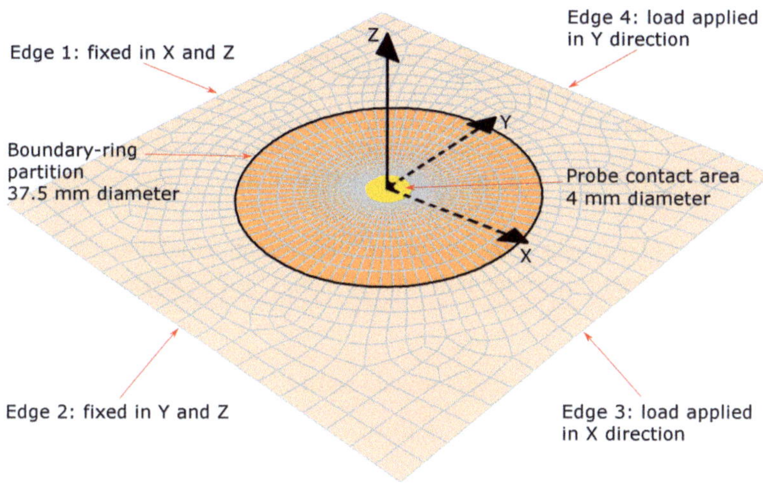

Fig. 5 Finite element model of the in vivo experiments

In an initial static analysis step, we applied a pre-stress (σ_x, σ_y) to the skin. Edge 1 was fixed in the X direction and edge 2 was fixed in the Y direction (Fig. 5). All four edges of the square were fixed in the Z direction. A load in the Y direction was applied to each node along edge 3 and a load in the X direction was applied to each node along edge 4. The loads were linearly increased from zero to the full load in 5 s.

In the second static analysis step, all the nodes outside the boundary-ring partition were fixed in all degrees of freedom. These conditions reflect the assumption that any skin outside the boundary ring did not move in the experiments. The nodes inside the probe area were displaced according to the in vivo experimental protocol. The total of the nodal reaction forces inside the probe region was calculated for the full cycle.

Constitutive Models

We investigated the suitability of three constitutive models to represent the mechanical behaviour of skin.

The orthotropic Bischoff et al. (2002) model is based on an assembly of eight fibres arranged within a unit cell. The orthotropic response is determined by the relative length of the cell edges *a*, *b* and *c*. This constitutive equation was implemented into FEBio using the user material plug-in facility. The strain energy function is

$$W_{\text{Bischoff}} = \frac{nk\theta}{4}\left(P^2\sum_{i=1}^{4}\left[\frac{\rho^{(i)}}{P^2}\beta_\rho^{(i)} + \ln\frac{\beta_\rho^{(i)}}{\sinh\beta_\rho^{(i)}}\right] - \frac{\beta_P}{P}\ln\left[\bar{\lambda}_{\mathbf{a}}^{a^2}\bar{\lambda}_{\mathbf{b}}^{b^2}\bar{\lambda}_{\mathbf{c}}^{c^2}\right]\right) + U(J) \quad (1)$$

where n is the number of fibres per unit volume, $k = 1.38 \times 10^{-23}\,\text{JK}^{-1}$ is Boltzman's constant, and θ is the absolute temperature.

$\bar{\lambda}_{\mathbf{a}} = \sqrt{\mathbf{a}^T\bar{\mathbf{C}}\mathbf{a}}$, $\bar{\lambda}_{\mathbf{b}} = \sqrt{\mathbf{b}^T\bar{\mathbf{C}}\mathbf{b}}$, $\bar{\lambda}_{\mathbf{c}} = \sqrt{\mathbf{c}^T\bar{\mathbf{C}}\mathbf{c}}$ are the principal fibre stretches along the principal material axes of the unit cell, $(\mathbf{a}, \mathbf{b}, \mathbf{c})$. $\bar{\mathbf{C}}$ is the deviatoric right Cauchy tensor.

$P = \frac{1}{2}\sqrt{a^2 + b^2 + c^2}$ is the undeformed length of a fibre in the cell, while $\rho^{(i)}$ is the deformed length of the ith fibre. $\beta_\rho^{(i)} = \mathcal{L}^{-1}\left(\frac{\rho^{(i)}}{N}\right)$, with $\mathcal{L}(x) = \coth x - \frac{1}{x}$ being the Langevin function.

The volumetric component of the strain energy function is

$$U(J) = \frac{B}{2}(\ln J)^2 \quad (2)$$

where $B = 1$ MPa is the bulk modulus and $J = det\mathbf{F}$ is the volume ratio. $\mathbf{F}$ is the deformation gradient.

The second model tested was a frame invariant version of the Fung constitutive equation (Ateshian and Costa 2009).

$$W_{\text{Fung}} = \frac{c}{2}\left(\mathrm{e}^{Q} - 1\right) + U(J) \quad (3)$$

$$Q = c^{-1}\sum_{a=1}^{3}\left[2\mu_a\mathbf{a}_a^0 \otimes \mathbf{a}_a^0 : \bar{\mathbf{E}}^2 + \sum_{b=1}^{3}\lambda_{ab}\left(\mathbf{a}_a^0 \otimes \mathbf{a}_a^0 : \bar{\mathbf{E}}\right)\left(\mathbf{a}_b^0 \otimes \mathbf{a}_b^0 : \bar{\mathbf{E}}\right)\right] \quad (4)$$

where c is a parameter representing the stiffness, and λ_{ab}, μ_a are Lamé parameters. All parameters have units of stress. $\bar{\mathbf{E}} = \frac{1}{2}(\bar{\mathbf{F}}^T\bar{\mathbf{F}} - \mathbf{I})$ is the deviatoric Green–Lagrange strain tensor. $\mathbf{a}_a^0$ defines an initial direction of a material axis a. For the purposes of simplifying the parameter optimisation procedure, $\lambda_{11} = \lambda_{12} = \lambda_{23} = \lambda_{31}$, $\lambda_{22} = \lambda_{33}$, and $\mu_2 = \mu_3$.

The third constitutive model tested was an anisotropic model proposed by Gasser et al. (2006).

$$W_{\text{Gasser}} = \frac{\mu}{2}(\bar{I}_1 - 1) + \frac{k_1}{k_2}\left\{\mathrm{e}^{k_2[\bar{I}_n(\theta)-1]^2} - 1\right\} + U(J) \quad (5)$$

where μ, k_1 control the stiffness of the skin at small strains and k_2 is a dimensionless parameter that controls the stiffness at large strains. $\bar{I}_n(\theta) = \mathbf{N}\bar{\mathbf{C}}\mathbf{N}$ is the fibre stretch squared of the nth family of fibres orientated in the direction $\mathbf{N}$ in the reference configuration.

For all the strain energy functions described, we used a quasi-linear viscoelastic model to characterise the time-dependent properties of skin (Fung 1993).

$$\mathbf{T}(t) = \mathbf{T}_e(t) + \int_0^t \mathbf{T}_e(t-\tau)\frac{\partial g_R(\tau)}{\partial \tau}\,\mathrm{d}\tau, \tag{6}$$

where $\mathbf{T}(t)$ is the total Cauchy stress at time t, $\mathbf{T}_e = \frac{1}{J}\mathbf{F}\frac{\partial W}{\partial E}\mathbf{F}^T$ is the elastic Cauchy stress, and $g_r(t)$ is a Prony series relaxation function.

$$g_R(t) = 1 - \bar{g}_1^P\left(1 - \mathrm{e}^{-t/\tau_1^G}\right) \tag{7}$$

where $\bar{g}_1^P = 0.4$ is a viscoelastic parameter and $\tau_1^G = 0.8\,s$ is the relaxation time.

Parameter Identification Framework

Using a framework developed in MATLAB 2016a (The MathWorks, Inc., Natick, MA, USA), we identified the material parameters and pre-stress field that best fit the probe reaction forces from the model to the measured probe reaction forces from the experiments. Optimised parameter sets using each constitutive model were determined.

Similar to Flynn et al. (2013), the following objective function was minimised in a least-squares sense using the *lsqnonlin* function in Matlab:

$$F(\mathbf{x}) = \sum_{i=1}^{16}\sum_{j=1}^{N_i}\left\{\left(\frac{R_{X_j}^{\mathrm{model}}(\mathbf{x}) - R_{X_j}^{\mathrm{exp}}}{\max\left(R_{X_j}^{\mathrm{exp}}\right)}\right)^2 + \left(\frac{R_{Y_j}^{\mathrm{model}}(\mathbf{x}) - R_{Y_j}^{\mathrm{exp}}}{\max\left(R_{Y_j}^{\mathrm{exp}}\right)}\right)^2 + \left(\frac{R_{Z_j}^{\mathrm{model}}(\mathbf{x}) - R_{Z_j}^{\mathrm{exp}}}{\max\left(R_{Z_j}^{\mathrm{exp}}\right)}\right)^2\right\} \tag{8}$$

where $\mathbf{x}$ is the model parameter set, and N_i is the number of data points recorded for the ith probe direction (16 directions in total). $R_{X_j}^{\mathrm{model}}(\mathbf{x})$, $R_{Y_j}^{\mathrm{model}}(\mathbf{x})$, and $R_{Z_j}^{\mathrm{model}}(\mathbf{x})$ are the model probe reaction forces in the X, Y, and Z directions at the jth data point. $R_{X_j}^{\mathrm{exp}}$, $R_{Y_j}^{\mathrm{exp}}$, $R_{Z_j}^{\mathrm{exp}}$ are the experiment probe reaction forces in the X, Y and Z directions at the jth data point.

The procedure of the MATLAB script for the optimisation was as follows. Starting with initial material parameters and a pre-stress field, a simplified mesh with no partitions was subjected to the boundary and load conditions specified in the first static step described in the Finite element model section. This step determined the stretch of the model in the X and Y directions. Using the stretch information, a Gmsh script created the full finite element mesh with partitions described in the Finite element model section. The geometries of the boundary and probe partitions were such that when the pre-stress was applied to this mesh in the initial step, both these partitions were circular and their diameters corresponded to those of the experiment. The second step with the nodes in the probe region being displaced was then executed.

When all analyses were complete, the objective function in Eq. (8) was calculated. The material parameters and pre-stresses were then adjusted by the *lsqnonlin* function. Updated FEBio and Gmsh input files were created and the analyses executed again. This iterative process continued until the objective function reached a local minimum.

We identified parameter sets that best fit the model data to in vivo data for different points of the face. For each optimised set, the variance accounted for (*VAF*) was calculated.

$$\mathrm{VAF} = 1 - \frac{F(\mathbf{x})}{\sum_{i=1}^{16}\sum_{j=1}^{N_i}\left\{\left(\frac{R_{X_j}^{\exp}}{\max\left(R_{X_j}^{\exp}\right)}\right)^2 + \left(\frac{R_{Y_j}^{\exp}}{\max\left(R_{Y_j}^{\exp}\right)}\right)^2 + \left(\frac{R_{Z_j}^{\exp}}{\max\left(R_{Z_j}^{\exp}\right)}\right)^2\right\}} \quad (9)$$

The results were compared with the results using the Ogden (1972) model reported in Flynn et al. (2013).

Results

Through the nonlinear optimisation procedure, we identified the material parameters and pre-stresses that best fit several constitutive models to in vivo experimental data (Tables 1, 2, 3 and 4). The lowest *VAF* was 79% for the Gasser et al. (2006) model when simulating the skin deformation in the forehead region Table 1. The

Table 1 Forehead region: identified material parameters, in vivo pre-stress field, and variance accounted for (*VAF*)

Model	Model parameters	(σ_x, σ_y) (kPa)	*VAF* (%)
Bischoff et al. (2002)	$n = 6.632 \times 10^{11}\mathrm{mm}^{-3}$; $(a, b, c) = (0.8529, 1.272, 1.386)$	(24.52, 22.73)	94
Ateshian and Costa (2009)	$c = 0.3118$ kPa; $\lambda_{11} = 0.9982$ kPa; $\lambda_{22} = 1.005$ kPa; $\mu_1 = 7.169$ kPa; $\mu_2 = 7.142$ kPa	(7.336, 3.351)	94
Gasser et al. (2006)	$\mu = 14.08$ kPa; $k_1 = 11.01$ kPa; $k_2 = 0.09188$; $\theta = 30.82^o$	(25.65, 17.67)	79
Ogden (1972)	$\mu_1 = 53.95$ kPa; $\mu_2 = 0.3012$ Pa; $\alpha_1 = 1.868$; $\alpha_2 = 69.00$	(34.12, 26.72)	94

The Ogden (1972) parameters are from Flynn et al. (2013)

Table 2 Near lip region: Identified material parameters, in vivo pre-stress field, and variance accounted for (*VAF*)

Model	Model parameters	(σ_x, σ_y) (kPa)	VAF (%)
Bischoff et al. (2002)	$n = 5.690 \times 10^{11}\text{mm}^{-3}$; $(a, b, c) = (0.6952, 1.276, 1.529)$	(9.088, 7.904)	94
Ateshian and Costa (2009)	$c = 0.3291$ kPa; $\lambda_{11} = 1.0$ kPa; $\lambda_{22} = 6.387$ kPa; $\mu_1 = 4.556$ kPa; $\mu_2 = 2.352$ kPa	(9.121, 7.342)	91
Ogden (1972)	$\mu_1 = 41.29$ kPa; $\mu_2 = 0.16$ Pa; $\alpha_1 = 1.658$; $\alpha_2 = 54.964$	(24.2, 15.9)	93

The Ogden (1972) parameters are from Flynn et al. (2013)

Table 3 Central cheek region: identified material parameters, in vivo pre-stress field, and variance accounted for (*VAF*)

Model	Model parameters	(σ_x, σ_y) (kPa)	*VAF* (%)
Bischoff et al. (2002)	$n = 1.246 \times 10^{12}\text{mm}^{-3}$; $(a, b, c) = (0.7880, 1.246, 1.446)$	(52.52, 45.98)	92
Ateshian and Costa (2009)	$c = 0.4421$ kPa; $\lambda_{11} = 1.0$ kPa; $\lambda_{22} = 1.0$ kPa; $\mu_1 = 6.121$ kPa; $\mu_2 = 4.353$ kPa	(9.518, 4.970)	93
Ogden (1972)	$\mu_1 = 58.27$ kPa; $\mu_2 = 0.14$ Pa; $\alpha_1 = 2.334$; $\alpha_2 = 33.081$	(89.4, 71.8)	93

The Ogden (1972) parameters are from Flynn et al. (2013)

Table 4 Central jaw region: identified material parameters, in vivo pre-stress field, and variance accounted for (*VAF*)

Model	Model parameters	(σ_x, σ_y) (kPa)	VAF (%)
Bischoff et al. (2002)	$n = 8.922 \times 10^{11}\text{mm}^{-3}$; $(a, b, c) = (0.8406, 1.224, 1.413)$	(36.88, 31.66)	96
Ateshian and Costa (2009)	$c = 0.5010$ kPa; $\lambda_{11} = 1.000$ kPa; $\lambda_{22} = 1.000$ kPa; $\mu_1 = 3.322$ kPa; $\mu_2 = 5.919$ kPa	(20.33, 15.84)	95
Ogden (1972)	$\mu_1 = 57.73$ kPa; $\mu_2 = 0.42$ Pa; $\alpha_1 = 2.265$; $\alpha_2 = 34.689$	(81.3, 75.4)	96

The Ogden (1972) parameters are from Flynn et al. (2013)

highest *VAF* was 96% for both the Bischoff et al. (2002) and Ogden (1972) models when simulating deformations in the central jaw region (Table 4).

The Bischoff et al. (2002) model simulated the nonlinear, anisotropic, and viscoelastic response of forehead skin observed in the experiments (Fig. 6). The results of the model for the other facial regions were similar. The Ateshian and Costa (2009)

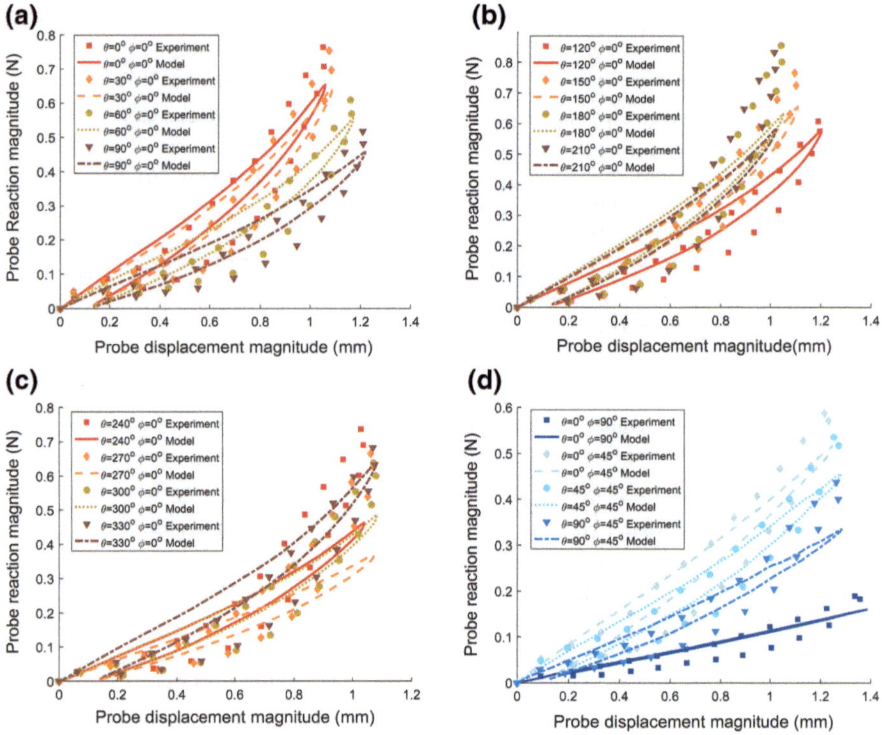

Fig. 6 Experiment and Bischoff et al. (2002) model probe reaction–displacement response for forehead region. **a–c** in-plane response; **d** out-of-plane response. *VAF* for Bischoff et al. (2002) was 94%. See Table 1 for model parameters and pre-stresses

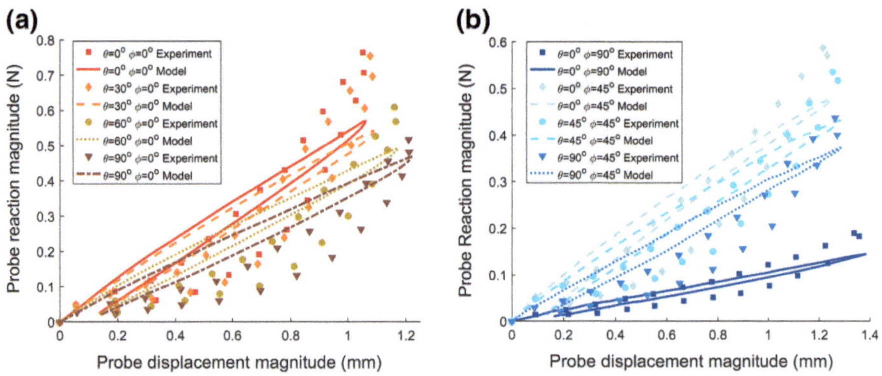

Fig. 7 Experiment and Gasser et al. (2006) model probe reaction–displacement response for forehead region. **a** sample of in-plane response; **b** out-of-plane response. *VAF* for Gasser et al. (2006) was 79%. See Table 1 for model parameters and pre-stresses

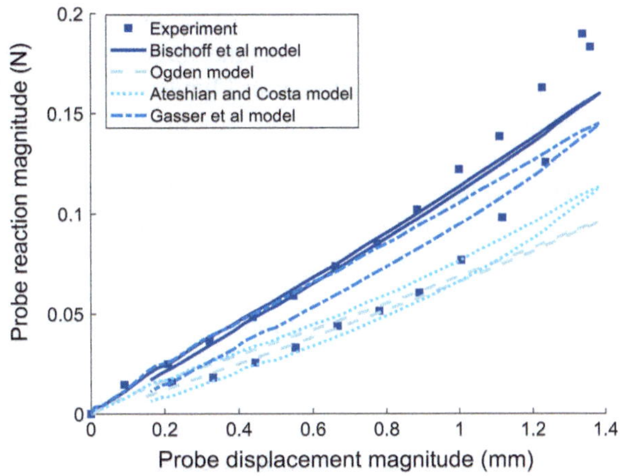

Fig. 8 Comparison of the forehead region normal response ($\theta = 0^{\circ}, \phi = 90^{\circ}$) of the models to the experimental measurement

model exhibited similar behaviour to the Bischoff et al. (2002) model (results not shown). The *VAF*s for both models in all regions were similar.

The Gasser et al. (2006) model did not simulate the nonlinear response of the forehead skin as well as the other models (Fig. 7).

The constitutive models simulated the normal deformation of the forehead region with varying degrees of accuracy (Fig. 8). The peak probe reaction force of the Ateshian and Costa (2009) model was approximately 50% of the measured force. The Bischoff et al. (2002) model simulated a maximum reaction force that was approximately 85% of the measured normal response.

Discussion

The force response of several facial regions undergoing a rich set of deformations has been simulated by a finite element model. Three anisotropic constitutive equations were tested and the *VAF* between the model and the experimental results ranged from 79% to 96%.

For each constitutive equation, the material parameter values varied considerably according to facial region. This demonstrates the need to measure the response at many points to develop a realistic facial model.

The *VAF*s using the Bischoff et al. (2002), Ateshian and Costa (2009) and Ogden (1972) models were very similar and differed by no more than a few per cent at each facial location. In contrast, the *VAF* for the Gasser et al. (2006) model was poor compared to the other models and its force–displacement response was too linear. Using anisotropic constitutive models did not yield better results than using the isotropic Ogden (1972) model. However, if knowledge of the collagen fibre distributions were obtained from appropriate imaging, this information could be used by the anisotropic models and could yield improved results.

In general, the force response of the model at small probe displacements was too high compared to the experimental force response. This contrasts with the Ogden (1972) model results reported in Flynn et al. (2013). In that study, there was better agreement between model and experiment at lower displacements than at higher displacements.

The estimated pre-stress in a region varied according to the constitutive model used. For example, in the forehead region, using the Ateshian and Costa (2009) model resulted in a pre-stress in the X direction of 7.336 kPa and 3.351 kPa in the Y direction Table 1. This was a lower pre-stress than estimated by the other models. When comparing the normal force–displacement responses, the Ateshian and Costa (2009) model had the poorest agreement with the experimental data (Fig. 8). Flynn et al. (2011) demonstrated that the calculated normal response of skin is more dependent on the pre-stress than the material parameters. This idea has been explored in calculating the wall stress in cerebral aneurysms (Lu et al. 2008). The estimated pre-stress using the Bischoff et al. (2002) model (24.52 kPa, 22.73 kPa) was similar to the pre-stress estimated using the Gasser et al. (2006) model (25.65 kPa, 17.67 kPa). Their respective normal force–displacement responses were also similar (Fig. 8). However, the estimated pre-stress using the Ogden (1972) model was the highest (34.12 kPa, 26.72 kPa) but the corresponding normal response did not match the experimental data. Further investigation is needed to determine the effect of the constitutive model on the estimated pre-stress and normal response.

There was some asymmetry in the measured force–displacement response. The simplistic nature of the model does not allow this asymmetry to be simulated. Improvements should be observed if the sub-dermal layers and their attachments are represented.

References

Ateshian GA, Costa KD (2009) A frame-invariant formulation of Fung elasticity. J Biomech 42 (6):781–785

Bischoff JE, Arruda EA, Grosh K (2002) A microstructurally based orthotropic hyperelastic constitutive law. J Appl Mech Trans ASME 69(5):570–579

Flynn C, Taberner A, Nielsen P (2011) Modeling the mechanical response of in vivo human skin under a rich set of deformations. Ann Biomed Eng 39(7):1935–1946

Flynn C, Taberner AJ, Nielsen PMF, Fels S (2013) Simulating the threedimensional deformation of in vivo facial skin. J Mech Behav Biomed Mater 28:484–494

Flynn C, Stavness I, Lloyd J, Fels S (2015) A finite element model of the face including an orthotropic skin model under in vivo tension. Comput Methods Biomech Biomed Eng 18 (6):571582

Fung YC (1993) Biomechanics: mechanical properties of living tissues. Springer, New York

Gasser TC, Ogden RW, Holzapfel GA (2006) Hyperelastic modelling of arterial layers with distributed collagen fibre orientations. J R Soc Interface 3(6):15–35

Geuzaine C, Remacle J-F (2009) Gmsh: A 3-D finite element mesh generator with built-in pre-and post-processing facilities. Int J Numer Methods Eng 79(11):1309–1331

Lu J, Zhou X, Raghavan ML (2008) Inverse method of stress analysis for cerebral aneurysms. Biomech Model Mechanobiol 7(6):477–486

Luboz V, Promayon E, Payan Y (2014) Linear elastic properties of the facial soft tissues using an aspiration device: towards patient specific characterization. Ann Biomed Eng 42(11):2369–2378

Maas SA, Ellis BJ, Ateshian GA, Weiss JA (2012) FEBio: finite elements for biomechanics. J Biomech Eng 134(1):011005

Ogden RW (1972) Large deformation isotropic elasticity—on the correlation of theory and experiment for incompressible rubberlike solids. Proc R Soc Lond A Math Phys Sci 326 (1567):565–584

Ohshima H, Tada A, Kanamaru A, Akamatsu H, Sakai Y, Itoh M, Kanto H (2011) Relevance of the directionality of skin elasticity to aging and sagging of the face. Skin Res Technol 17(1):101–107

Swain D, Gupta A (2015) Interfacial growth during closure of a cutaneous wound: stress generation and wrinkle formation. Soft Matter 11(32):6499–6508

Weickenmeier J, Jabareen M, Mazza E (2015) Suction based mechanical characterization of superficial facial soft tissues. J Biomech 48(16):4279–4286

Wu T-F, Hung A, Mithraratne K (2014) Generating facial expressions using an anatomically accurate biomechanical model. IEEE Trans Vis Comput Graph 20(11):1519–1529

Zhang X, Tang Z, Liebschner MA, Kim D, Shen S, Chang C-M, Yuan P, Zhang G, Gateno J, Zhou X et al (2016) An eFace-template method for efficiently generating patient-specific anatomically-detailed facial soft tissue FE models for craniomaxillofacial surgery simulation. Ann Biomed Eng 44(5):1656–1671

Evaluation of a Mechanically Coupled Reaction–Diffusion Model for Macroscopic Brain Tumor Growth

Daniel Abler and Philippe Büchler

Abstract The macroscopic growth of brain tumors has been studied by means of different computational modeling approaches. Glioblastoma multiforme (GBM) is the most common malignant type and is commonly modeled as a reaction–diffusion type system, accounting for its invasive growth pattern. Purely biomechanical models have been proposed to represent the mass effect caused by the growing tumor, but only a few models consider mass effect and tissue invasion effects in a single 3D model. We report first results of a comparative study that evaluates the ability of a simple computational model to reproduce the shape of pathologies found in patients. GBM invasion into brain tissue and the mechanical interaction between tumor and healthy tissue components are simulated using the finite element method (FEM). Cell proliferation and invasion are modeled as a reaction–diffusion process; simulation of the mechanic interaction relies on a linear elastic material model. Both are coupled by relating the local increase in tumor cell concentration to the generation of isotropic strain in the corresponding tissue element. The model accounts for multiple brain regions with values for proliferation, isotropic diffusion, and mechanical properties derived from literature. Tumors were seeded at multiple locations in FEM models derived from publicly available human brain atlases. Simulation results for a given tumor volume were compared to patient images. Simulated tumors showed a more symmetric growth pattern compared to their real counterparts. Resulting levels of tumor invasiveness were in agreement with simulation parameters and tumor-induced pressures of realistic magnitude were found.

D. Abler (✉) · P. Büchler
Institute for Surgical Technology and Biomechanics (ISTB),
University of Bern, Bern, Switzerland
e-mail: daniel.abler@istb.unibe.ch

P. Büchler
e-mail: philippe.buechler@istb.unibe.ch

A. Gefen and D. Weihs (eds.), *Computer Methods in Biomechanics and Biomedical Engineering*, Lecture Notes in Bioengineering,
DOI 10.1007/978-3-319-59764-5_7

Introduction

Brain tumors represent a rare but serious medical condition. With an incidence of six cases per 100,000, gliomas are the most frequent primary brain tumors in adults, accounting for 70% (Ricard et al. 2012) of cases. Gliomas are classified into four grades by increasing aggressiveness, based on their microscopic structure and cellular activity. Glioblastoma multiforme (GBM) is the most frequent and most malignant subtype of glioma (grade IV), accounting for about 50% of diffuse gliomas. These tumors infiltrate surrounding healthy tissue, grow rapidly, and form a necrotic core of high cell density which is often accompanied by compression and displacement of surrounding tissue. This so-called 'mass effect' leads to an increase in intracranial pressure (ICP) and the progressive onset of a multitude of pressure-related symptoms, from headache and nausea to coma or death due to herniation which is the leading cause of death for GBM patients (Sizoo et al. 2010). The standard treatment therefore involves surgical resection of the bulk tumor to reduce the symptoms of mass effect, followed by a combination of chemo- and/or radiation therapy. Long-term prognosis for GBM remains poor, with median overall survival between 1 and 2 years (Ricard et al. 2012).

Different types of mathematical models (discrete, continuous, hybrid) on different spatial and spatiotemporal scales have been employed to improve the understanding of GBM and to optimize treatment approaches (Hatzikirou et al. 2005). Continuous representations are particularly well suited for clinical applications as they allow modeling the temporal evolution of macroscopic processes on the tissue level and at low computational cost. Emphasis has been placed on simulating the invasion dynamics of glioma due to the direct clinical importance of this growth characteristic. A framework for simulating GBM invasion was introduced by Tracqui et al. (1995), Woodward et al. (1996), Burgess et al. (1997) based on a reaction–diffusion equation that describes tumor cell migration as a random walk process modeled by Fickian diffusion. These initial models were soon extended to account for more realistic brain geometries and differences between cell motility in gray and white matter (Swanson et al. 2000). Patient-specific parameters have been estimated from routine clinical imaging to compute the invisible portion of the tumor and to predict tumor recurrence after surgical resection (Swanson et al. 2003). Further extensions aim to incorporate heterogeneity and anisotropy of the brain anatomy informed by routine clinical MRIs (Jbabdi et al. 2005) and consider the effects of different treatments (Swanson et al. 2002, 2008a; Rockne et al. 2009).

Despite the recognized importance of the biomechanical environment for tumor evolution (Jain et al. 2014), the mass effect caused by the growing tumor received less attention from the modeling community and has been investigated mainly in the context of improving image registration methods (Mohamed and Davatzikos 2005; Hogea et al. 2007; Gooya et al. 2012; Bauer et al. 2012) for atlas-based segmentation. Few models, such as (Clatz et al. 2005; Hogea et al. 2008), consider both tumor invasion and mass effect in a realistic 3D model of the human brain using solid mechanics approaches. Accurate morphological representation is crucial also for the validation of macroscopic tumor models, which often relies on multi-modality imaging data. Indeed, the lack of anatomical correspondence has been identified as one of the limitations of advanced reaction–diffusion based GBM models (Gu et al. 2012).

We report first results of a comparative study that evaluates the ability of a simple computational model of mechanically coupled diffusive tumor growth to reproduce characteristics of pathologies found in patients. GBM invasion into brain tissue and the mechanical interaction between tumor and healthy tissue components are simulated using the finite element method (FEM).

Materials and Methods

The mathematical model employed in this study is designed to capture three interrelated aspects of macroscopic glioma growth: (a) tumor cell proliferation, (b) the diffuse invasion of the growing tumor into surrounding healthy tissue, and (c) the resulting mass effect. The model operates on a domain that represents the human brain. For the present study, each component of the domain was considered homogeneous, with distinct isotropic material properties in each subdomain: white matter (WM), gray matter (GM) and cerebrospinal fluid (CSF).

The invasive growth pattern of glioma was modeled as a reaction–diffusion process (Tracqui et al. 1995; Woodward et al. 1996; Burgess et al. 1997):

$$\frac{\partial c}{\partial t} = \nabla \cdot (D \nabla c) + \rho c(1 - c),$$

where c represents the normalized concentration of tumor cells in function of time and space. The isotropic and locally constant diffusion D represents the migration rate of GBM cells in brain tissue, with higher values in WM than in GM (Swanson et al. 2000). Logistic growth with growth rate ρ was assumed as reaction term. This formulation is known as Fisher's equation (Fisher 1937) and leads to the creation of a non-proliferating zone in regions of high tumor cell concentration, in agreement with clinical observation of the formation of a necrotic core.

The creation of new tumor cells during growth results in volumetric increase of the tumor and introduces growth-related strain ε^g in the tissue. The present model assumes a linear relation between local cell concentration and tumor-induced strain with a coupling constant λ:

$$\hat{\epsilon}^g = \lambda c$$

Simulations were carried out over the SRI24 human brain atlas (Rohlfing et al. 2010). This atlas represents the normal brain anatomy and provides separate labels for WM, GM, and CSF. The CSF domain was subdivided into two compartments, separating the fluid-filled brain ventricles from the remaining CSF. Brain tissues were

Table 1 Mechanical parameters for model constituents

Tissue	E (kPa)	ν
White/gray matter	3.0	0.45
Tumor	6.0	0.45
CSF (ventricles)	1.0	0.30
CSF (other)	1.0	0.49

The CSF of the ventricles is modeled as compressible elastic solid to account for physiological compensation mechanisms

Table 2 Reaction–diffusion parameters in GM and WM for different levels of invasiveness

Invasiveness $\sim D/\rho$	ρ (day^{-1})	D_{GM} (mm^2 day^{-1})	D_{WM} (mm^2 day^{-1})
Low	0.082	0.020	0.101
Medium	0.046	0.022	0.110
High	0.037	0.040	0.200

Zero diffusion D and proliferation ρ are assumed throughout both CSF compartments

modeled as linear elastic materials in this study with Poisson ratio ν and Young's modulus E as the mechanical properties. The CSF of the brain ventricles was modeled as a compressible elastic solid to account for physiological mechanisms that compensate elevated ICP. Mechanical parameters for the model constituents are chosen similar to Wittek et al. (2010) and are summarized in Table 1.

Parameter choices for diffusivity and proliferation were derived from literature data based on clinical observations in glioblastoma patients (Swanson et al. 2008b; Wang et al. 2009; Szeto et al. 2009; Rockne et al. 2010). Under the assumption of fivefold higher diffusion in WM compared to GM (Swanson et al. 2000), three sets of parameters ($D_{GM/WM}$, ρ) were obtained, corresponding to low, medium, and high levels of invasiveness, Table 2.

A volumetric expansion coefficient $\lambda = 0.15$ was chosen for all simulations, corresponding to a maximum tumor-related increase in local volume of 15% (Bauer et al. 2012).

The healthy brain atlas was seeded at multiple locations, and simulated until the tumor had reached a maximum volume corresponding to 3.5 cm equivalent radius. Seed positions were obtained from a publicly accessible dataset of preoperative GBM images (NCI–MICCAI 2013), which were registered to the healthy atlas and segmented using an automatic GBM segmentation software (Porz et al 2014). Center-of-mass positions of the tumors visible in T1-weighted MRI were extracted and used as patient-specific seed position for all patients of the dataset, Fig. 1.

A tetrahedral mesh was generated from each seeded atlas using CGAL (http://www.cgal.org). Spatial confinement of the brain within the skull was modeled by imposing zero-displacement and zero-flux boundary conditions on the surface nodes of the mesh. The model was implemented as thermal expansion analysis in Abaqus (Simulia) by identifying relative cell concentration c with 'temperature' as simulation variable.

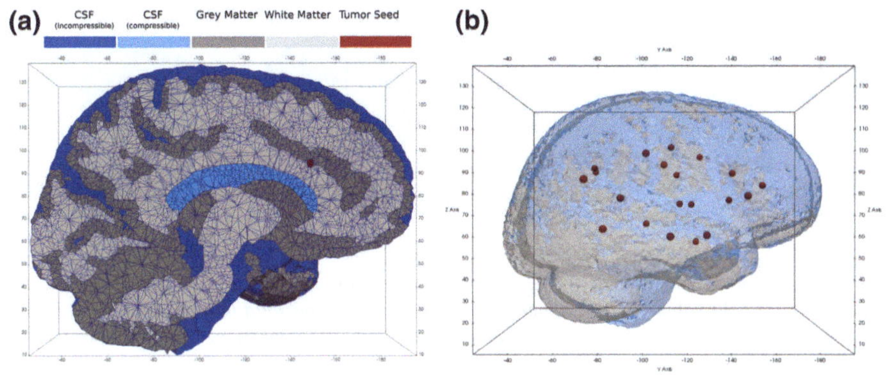

Fig. 1 Cross-section of the anatomical atlas for an exemplary seeding scenario (**a**), and overview of all seed positions within the healthy brain atlas (**b**)

For each seed position, three tumor growth simulations were carried out, corresponding to the different levels of invasiveness defined in Table 2. Simulated tumors were compared to actual tumors in terms of shape and invasiveness when volumetric agreement was reached. Additionally, the mechanical impact of the simulated tumor was computed.

Results

Simulation of this mechanically coupled reaction–diffusion model yields as outputs the temporal evolution of tumor cell concentration over the brain anatomy as well as the mechanical impact of the tumor in terms of displacement and tissue stresses.

Parameter choices for different levels of invasiveness were correctly reflected in simulation results with larger D/ρ consistently resulting in a higher portion of invisible ($c < 0.8$) compared to visible ($c > 0.8$) tumor.

While the shape of near-spherical tumors was well reproduced in simulations, more complex shapes were not. Figure 2 compares simulation and actual tumor for two representative tumor shapes. Statistical evaluation showed the simulated tumors to be more symmetric, as measured by their aspect ratio, compared to their real counterparts.

The mechanical impact of the growing tumor was estimated for each simulated case by computing the tumor-induced pressure acting on the skull. For a fatal tumor burden, assumed at 3.5 cm equivalent radius (Swanson et al. 2008a), simulations yield excess pressures between 1.0 and 1.4 kPa in addition to the normal intracranial pressure of about 0.7–2.0 kPa (Rangel-Castillo et al. 2008).

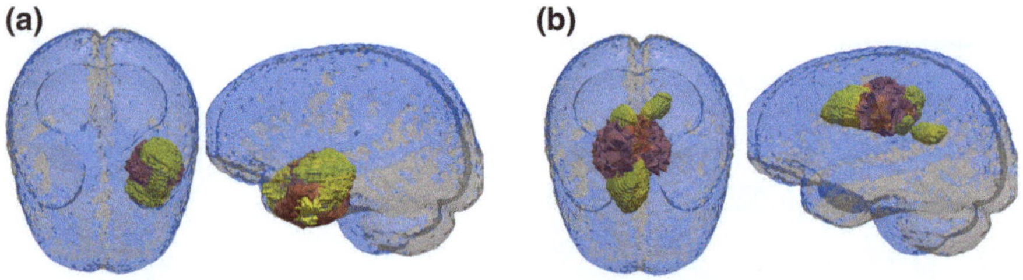

Fig. 2 Comparison of simulated (*red*) to observed tumor shapes (*yellow*) for selected cases. Concentration field values $c \geq 0.8$ were chosen to represent the *solid tumor*. Good agreement is observed for regularly structured shapes (**a**), whereas highly irregular shapes are not reproduced in simulations (**b**)

Discussion

The tumor model in this implementation assumed isotropic and locally constant material properties and did not take into account biological processes other than cell proliferation. Growth and mechanical properties were not personalized and simulations were carried out over a standardized atlas of healthy brain anatomy. These limitations explain the observed discrepancy in the shapes of simulated compared to real tumors from the dataset, particularly the overall underestimation of tumor asymmetry in simulations. Inclusion of anisotropic diffusion properties, with increased diffusion along the fiber structure of the brain, has been shown to improve spatial agreement between simulated and actual tumor (Jbabdi et al. 2005) in a pure reaction–diffusion model. Anisotropic diffusion, as well as anisotropic mechanical material properties, can be integrated in this model.

Despite its simplicity, the model yielded realistic order-of-magnitude estimates of the mechanical impact of a growing tumor. According to simulations and under the assumption of linear elasticity, the excess pressure caused by a tumor of 3.5 cm equivalent radius would increase total ICP to up to 3.4 kPa representing a state of mild intracranial hypertension that requires treatment in most circumstances (Rangel-Castillo et al. 2008). Replacement of the currently linear elastic by a more accurate material model for brain tissue is expected to increase these estimates further into the range of life-threatening intracranial hypertension (>5.3 kPa; Rangel-Castillo et al. 2008).

Tumor seed positions were derived from center-of-mass positions of preoperative MR images. This approach is expected to provide realistic estimates under (mostly) isotropic growth conditions. However, simulation results indicate that the assumption of isotropic growth does not hold for irregularly shaped tumors, suggesting that center-of-mass position is not a suitable indicator for tumor seed location in those cases.

Conclusion

A FEM-based approach for simulating diffusive growth and mass effect of GBM was presented. Tumor evolution was simulated over a healthy brain atlas for a range of seed positions and three sets of reaction–diffusion parameters. We showed qualitative agreement of resulting tumor invasiveness with simulation parameters and found tumor-induced pressures of realistic magnitude. Comparison to real tumor shapes confirmed previous observations from a pure reaction–diffusion model (Jbabdi et al. 2005) that tumor shape depends on seed position and that asymmetric shapes cannot be reproduced by isotropic growth assumptions. We therefore plan to carry out a sensitivity study to investigate the role of seed position on shape and to extend the present model to account for anisotropic material properties.

Acknowledgement The research leading to these results has received funding from the European Union Seventh Framework Programme (FP7/2007-2013) under grant agreement n° 600841.

References

Bauer S et al (2012) Multiscale modeling for image analysis of brain tumor studies. IEEE Trans Biomed Eng 59(1):25–29
Burgess PK et al (1997) The interaction of growth rates and diffusion coefficients in a three-dimensional mathematical model of gliomas. J Neuropathol Exp Neurol 56(6):704–713
Clatz O et al (2005) Realistic simulation of the 3-D growth of brain tumors in MR images coupling diffusion with biomechanical deformation. IEEE Trans Med Imaging 24(10):1334–1346
Fisher RA (1937) The wave of advance of advantageous genes. A Eug 7(4):355–369
CGAL, Computational geometry algorithms library, http://www.cgal.org
Gooya A et al (2012) GLISTR: glioma image segmentation and registration. IEEE Trans Med Imaging 31(10):1941–1954
Gu S et al (2012) Applying a patient-specific bio-mathematical model of glioma growth to develop virtual [18F]-FMISO-PET images. Math Med Biol 29(1):31–48
Hatzikirou H et al (2005) Mathematical modelling of glioblastoma tumour development: a review. Math Models Methods Appl Sci 15(11):1779–1794
Hogea C et al (2007) Modeling glioma growth and mass effect in 3D MR images of the brain. MICCAI, Springer, pp 642–650
Hogea C et al (2008) An image-driven parameter estimation problem for a reaction–diffusion glioma growth model with mass effects. J Math Biol 56(6):793–825
Jain R et al (2014) The role of mechanical forces in tumor growth and therapy. Ann Rev Biomed Eng 16(1):321–346
Jbabdi S et al (2005) Simulation of anisotropic growth of low-grade gliomas using diffusion tensor imaging. Magn Reson Med 54(3):616–624
Mohamed A, Davatzikos C (2005) Finite element modeling of brain tumor mass-effect from 3D medical images. MICCAI, Springer, pp 400–408
NCI–MICCAI (2013) Challenge on multimodal brain tumor segmentation. http://martinos.org/qtim/miccai2013/index.html
Porz et al (2014) Multi-modal glioblastoma segmentation: man versus machine. PLoS ONE 9(5): e96873. http://journals.plos.org/plosone/article/metrics?id=10.1371/journal.pone.0096873
Rangel-Castillo L et al (2008) Management of intracranial hypertension. Neurol Clin 26(2): 521–541

Ricard D et al (2012) Primary brain tumours in adults. Lancet 379(9830):1984–1996
Rockne R et al (2009) A mathematical model for brain tumor response to radiation therapy. J Math Biol 58(4–5):561–578
Rockne R et al (2010) Predicting the efficacy of radiotherapy in individual glioblastoma patients in vivo: a mathematical modeling approach. Phys Med Biol 55(12):3271–3285
Rohlfing T et al (2010) The SRI24 multi-channel atlas of normal adult human brain structure. Human Brain Mapp 31(5):798–819
Sizoo EM et al (2010) Symptoms and problems in the end-of-life phase of high-grade glioma patients. Neuro Oncol 12(11):1162–1166
Swanson KR et al (2000) A quantitative model for differential motility of gliomas in grey and white matter. Cell Prolif 33(5):317–329
Swanson KR et al (2002) Virtual brain tumours (gliomas) enhance the reality of medical imaging and highlight inadequacies of current therapy. Br J Cancer 86(1):14–18
Swanson KR et al (2003) Virtual resection of gliomas: effect of extent of resection on recurrence. Math Comput Model 37(11):1177–1190
Swanson KR et al (2008a) Velocity of radial expansion of contrast-enhancing gliomas and the effectiveness of radiotherapy in individual patients: a proof of principle. Clin Oncol 20(4): 301–308
Swanson KR et al (2008b) A mathematical modelling tool for predicting survival of individual patients following resection of glioblastoma: a proof of principle. Br J Cancer 98(1):113–119
Szeto MD et al (2009) Quantitative metrics of net proliferation and invasion link biological aggressiveness assessed by MRI with hypoxia assessed by FMISO-PET in newly diagnosed glioblastomas. Cancer Res 69(10):4502–4509
Tracqui P et al (1995) A mathematical model of glioma growth: the effect of chemotherapy on spatio-temporal growth. Cell Prolif 28(1):17–31
Wang CH et al (2009) Prognostic significance of growth kinetics in newly diagnosed glioblastomas revealed by combining serial imaging with a novel biomathematical model. Cancer Res 69(23):9133–9140
Wittek A et al (2010) Patient-specific non-linear finite element modelling for predicting soft organ deformation in real-time; application to non-rigid neuroimage registration. Prog Biophys Mol Biol 103:292–303
Woodward DE et al (1996) A mathematical model of glioma growth: the effect of extent of surgical resection. Cell Prolif 29(6):269–288

UKA Component Fatigue Test Development Using DOE and FEA

D. Levine, Y. Son, J. Phillips and J. Bischoff

Abstract Unicompartmental knee arthroplasty (UKA) is a clinical option for knee osteoarthritis continuum of care, providing a soft tissue sparing and less invasive approach compared to total knee arthroplasty (TKA). Compared to TKA, a unicompartmental procedure involves reduced blood loss, faster recovery, and fewer clinical complications. When designing new UKA implants, clinical risks must be assessed and mitigated. Implant fracture, while rare clinically, is an important consideration in design, and while design for minimal strength risk can be supported by simulation techniques and physical testing, published guidelines for such evaluations are not currently available. This publication presents a methodology developed in the course of verifying a new UKA tibial baseplate design and combines finite element analysis (FEA), design of experiments (DOE) and bench-top fatigue testing. The FE model and fatigue test setup have a three-point bend configuration which creates peak stress in the region where clinical fractures have occurred. Both FE model and fatigue test leveraged a draft ASTM standard for three-point bend testing of UKA baseplates. FEA and DOE were used in concert to identify a position for applied loading that produces worst-case stress in the implant. Subsequently, fatigue testing was performed on the implant that was analyzed by FEA and it was confirmed that the simulation accurately predicted the fatigue crack initiation site in implants tested to fracture. In recent years, the use of computational modeling to predict clinical performance has drawn the attention of regulatory agencies worldwide, and there is a growing expectation that such modeling must meet certain standards for verification and validation for it to be credible and also applicable to the performance questions it is used to answer. In this case, DOE explores the design environment by using not one but a range of

D. Levine (✉) · Y. Son · J. Phillips · J. Bischoff
Zimmer Biomet, Warsaw, IN, USA
e-mail: danny.levine@zimmerbiomet.com

A. Gefen and D. Weihs (eds.), *Computer Methods in Biomechanics and Biomedical Engineering*, Lecture Notes in Bioengineering,
DOI 10.1007/978-3-319-59764-5_8

load positions to find a worst-case scenario and to illustrate the sensitivity to model inputs. The bench-top test serves as a validation comparator, showing that the simulation is predictive of the product's physical behavior.

Introduction

Unicompartmental knee arthroplasty (UKA) is a clinical option for knee osteoarthritis continuum of care, providing a soft tissue sparing and less invasive approach compared to total knee arthroplasty (TKA). Compared to TKA, a unicompartmental procedure involves reduced blood loss, faster recovery, and fewer clinical complications. By preserving the cruciate ligaments, the procedure results in increased range of motion (Laurencin et al. 1991), improved joint kinematics (Patil et al. 2005), and a more natural patient "feel."

However; in order to achieve optimum clinical performance, careful surgical planning and skill are required to position and align the components (Sarmah et al. 2012). Clinical complications that have been observed over time include loosening and subsidence (O'Donnell et al. 2013) and implant component fracture (example shown in Fig. 1) (Palumbo et al. 2011).

While clinical fractures of UKA components are rare, their past occurrence highlights this potential risk as one that must be addressed by developers of new unicompartmental implant designs.

The present study is directed toward the management of risk associated with fatigue strength of UKA tibial baseplates, through a combination of finite element analysis (FEA) and bench-top fatigue testing. The FEA and bench-top test both utilize a simplified three-point bend configuration recently proposed by an ASTM

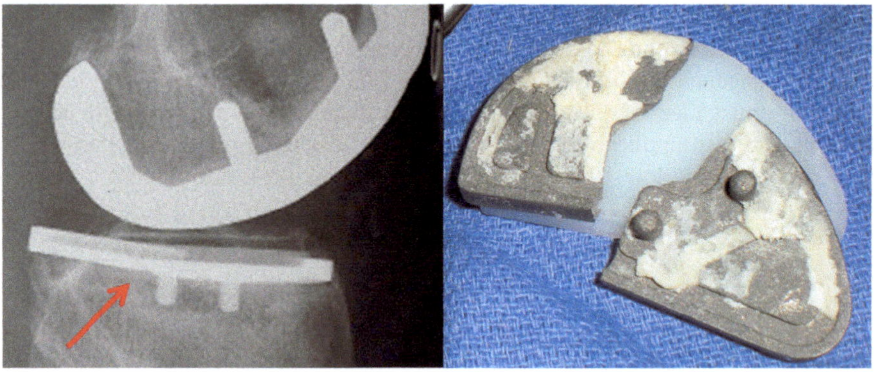

Fig. 1 Fractured UKA tibial baseplate (Sarmah et al. 2012)

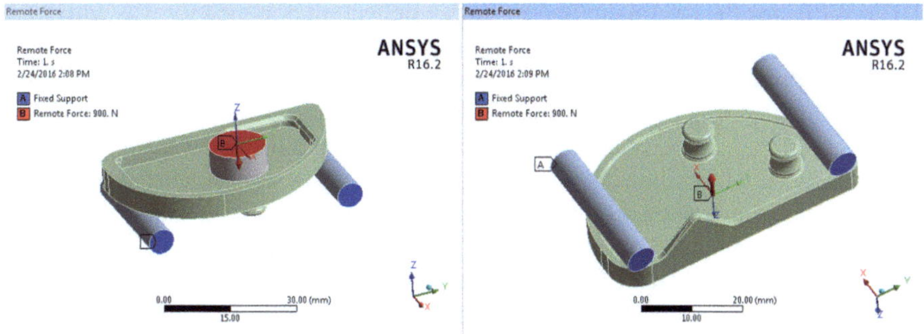

Fig. 2 Three-point bend configuration for FEA and test

working group (2016). While the proposed three-point bend configuration (Fig. 2) can produce fractures at clinically relevant locations, it lacks specific guidance for comparison of the worst-case condition experienced by a new design to that experienced by clinically available predicate; thus, it leaves to the end user the responsibility to define a fair basis for comparative evaluation. Since there are no published strength requirements for knee implants, comparison of a proposed design to an established predicate is an established approach for verifying safety of that new design.

A cursory examination of unicompartmental baseplates such as the one shown in Fig. 2 might prompt the use of simple beam theory to define the worst-case loading position which would, in general, mean that the A/P position of a worst-case load would be half way between the rollers. However; the majority of such baseplates incorporate attachment features such as pegs or keels on the inferior surface and these features can be stress risers. Therefore, the position of the applied load relative to these pegs, keels, etc., must be considered when defining the worst-case loading position.

The present work combines design of experiments (DOE) tools with finite element analysis (FEA) to identify worst-case loading for UKA tibial baseplates.

Materials and Methods

Models and results presented here illustrate the technique used for identifying the location of an applied load that produces worst-case stress in a UKA tibial baseplate and also provides a qualitative comparison of fatigue test results to FEA predictions.

All FE simulations were performed using the commercial software package ANSYS, Revision 16 (ANSYS, Inc., Canonsburg, PA, USA). Linearly elastic material properties were assigned to all components (Steel for the rollers, Ti-6Al-4 V for the baseplate and ultra-high molecular weight polyethylene

[UHMWPE] for the loading cylinder). Large deflection nonlinearity was active and contact nonlinearity was assumed at all interfaces.

For the predicate tibial baseplate presented here, CAD geometry for a Repicci II® Unicondylar baseplate (Biomet, Warsaw, Indiana, USA) was imported into ANSYS, and the geometry for the supporting rollers and the small loading cylinder were generated within the ANSYS environment using DesignModeler. The loading cylinder was constructed using a local coordinate system with anterior/posterior (A/P) and medial/lateral (M/L) coordinates of the origin set as parameters and subsequently used for DOE inputs. The range of A/P and M/L coordinates assigned to the center point of the loading cylinder is highlighted by the faint rectangle in Fig. 3. The A/P limits were based upon a published study (Akizuki et al. 2009) of unicompartmental knee kinematics. M/L limits were based upon model geometry. For the left-medial implant application illustrated by Fig. 3, the loading cylinder placement is limited on the left (toward the straight side) by a superior rail that constrains the articular surface component. On the right (toward the curved side), the cylinder placement is constrained by the roller support limits. If the load is too far to the right, the baseplate is subject to tipping.

The DOE study was performed using ANSYS DesignXplorer. A series of models using different values for the two input parameters (X and Y coordinates of the load cylinder) was generated using the proprietary optimal space filling algorithm which is similar to the Latin hypercube method. Following the solution of the

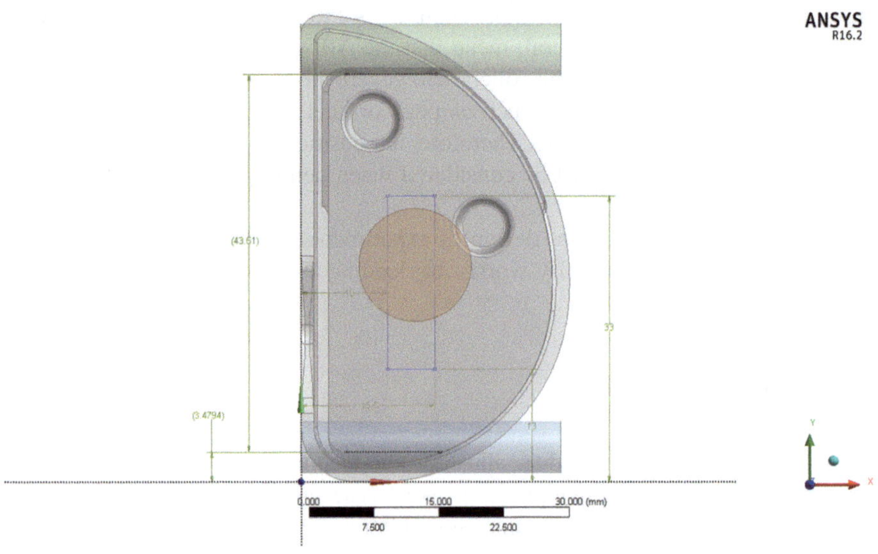

Fig. 3 Schematic of load and support placement used for FEA. Coordinates of the loading cylinder center point are referenced to the global coordinate system shown at the lower left corner of the figure

Fig. 4 Fatigue test configuration for Repicci Unicondylar Knee baseplate

set of models, resulting peak values of maximum principal stress were fit to a response surface using a Kriging algorithm with a 3% relative error. The Kriging algorithm ran additional parameter sets to generate surface refinement points to improve response surface smoothness and accuracy.

Following the guidance from ASTM draft WK45235 (2016), but using the worst-case load position obtained from the FE-based DOE study, fatigue testing was performed on samples (n = 10) of the same Repicci baseplate, as illustrated in Fig. 4. Fatigue testing was performed using a load ratio $R = 0.1$. Tests were performed at a number of different load levels in order to estimate the 10^7 cycle fatigue limit.

Results

Figure 5 presents the response surface generated from the models solved in the Repicci baseplate DOE. Individual result points are highlighted by open squares on the plot, and the response surface is smoothed to fit. The response surface shows local maxima and minima, but the overall maximum stress occurs when the load is close to the medial and anterior extremes of the allowed range.

Figure 6 shows the predicted maximum principal stress for the worst-case load position. The highest stress for this load case occurs at the anterior side of one fixation post.

When the corresponding bench-top fatigue test was performed, 8 of 10 specimens were tested to fracture. The fracture initiation site was consistent with the FE prediction, as illustrated by Fig. 7.

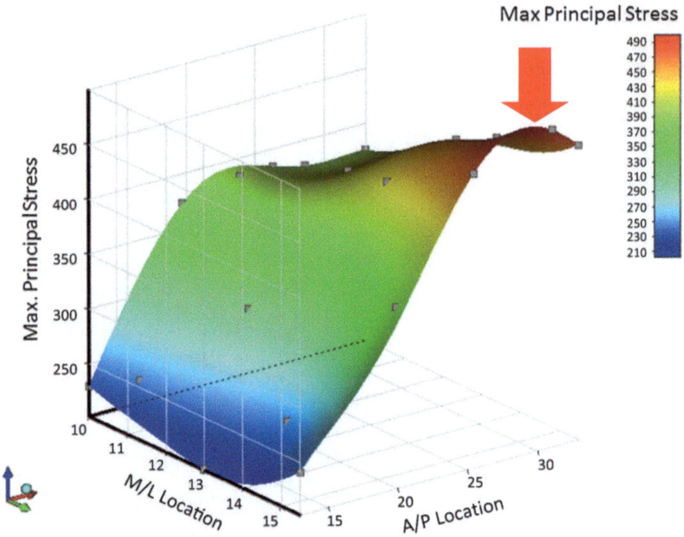

Fig. 5 Response surface for Repicci baseplate, showing peak Maximum Principal Stress (MPa) as a function of M/L and A/P load coordinates. The *arrow* indicates the overall highest stress predicted in the DOE simulation

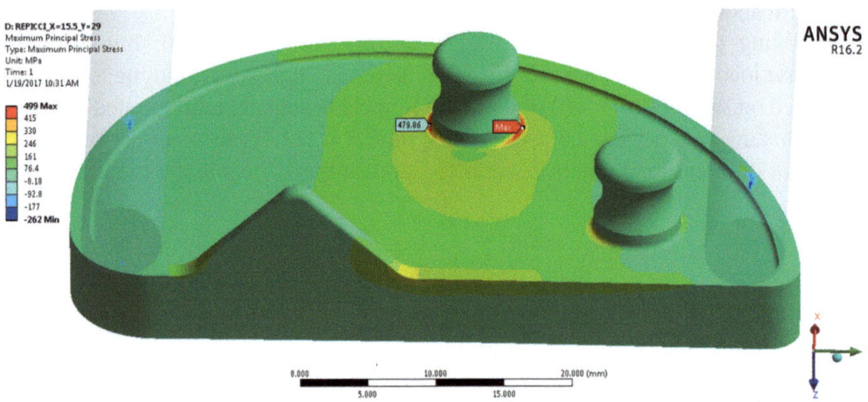

Fig. 6 Maximum principal stress (MPa) predicted for the worst-case load position

Discussion

The primary focus of this project was to define a fair means of comparing the performance of one product design to another with the same indications for use and similar features. The aim was to define an objective measure for worst-case loading

Fig. 7 Repicci baseplate fatigue specimen showing complete fracture (initiation region *highlighted*)

and to predict peak stress for that loading in a clinically available predicate device. That objective was successfully met.

There is more to making a fair (apples-to-apples) comparison than the specific performance metric that has been chosen. There is also, within the simulation community, an expectation that simulation best practices will be utilized for model quality assurance, and that modeling verification and validation (V&V) will be performed at a level commensurate with the risk associated with the decisions that the model informs.

For the medical device modeling and simulation community, ASME's V&V 40 subcommittee (https://cstools.asme.org/csconnect/CommitteePages.cfm?Committee=100108782) is leading the way in developing a framework that analysts (and experimentalists) may use to assess the risk associated with model-based decisions and to identify the V&V tasks and the level of effort applied to each task so as to manage the risk involved. The guidance document being created by this subcommittee is still in the writing stage, but the interested reader is directed to an existing publication, NASA-STD-7009 (2008), for an overview of V&V concepts and terminology.

Risk assessment touches upon two factors. The first factor is the consequence of an incorrect decision being made from model results. In this case, an incorrect design decision pertaining to tibial baseplate fatigue strength could lead to implant fracture, requiring revision surgery. The second factor is the influence which modeling has on the decision(s) being made. When modeling is used alone, with no testing, predicate or clinical history, the influence is high. However; this study includes both bench-top testing and a predicate, thereby lessening the model influence.

To ensure model credibility commensurate with the risk associated with the consequence and influence noted above, some of the pertinent V&V activities are as follows:

- Verification
  - Use of commercial off-the-shelf (COTS) software with established software development practices including code verification testing
  - Mesh convergence study to reduce discretization error
- Validation
  - Physical testing using the same configuration as FEA models
  - "Attribute" comparison between FEA prediction and test results [Either the tested part DOES or DOES NOT fail at the location predicted by simulation]

These V&V activities each confer a level of trust (credibility) in the models and the decisions being informed.

Methods for scoring risk and credibility are outside the scope of this discussion. The ASME V&V 40 working group is not intending to be prescriptive in this respect so it is up to the simulation users to develop a method for documenting computational risk assessment and credibility activities. For an example, the reader is referred to NASA-STD-7009 (2008).

References

Akizuki S et al (2009) In vivo determination of kinematics for subjects having a zimmer unicompartmental high flex knee system. J Arthroplasty 24(6):963–971

DRAFT (2016) Standard test method for cyclic fatigue testing of metal tibial tray components of unicondylar knee joint replacements. ASTM Draft WK45235. Contact: Gokce.Yildirim@Stryker.com

Laurencin CT, Zelicof SB, Scott RD et al (1991) Unicompartmental versus total knee arthroplasty in the same patient. A comparative study. Clin Orthop Relat Res 273:151

National Aeronautics and Space Administration Standard for Models and Simulations, NASA-STD-7009 (2008) NASA website. https://standards.nasa.gov/standard/nasa/nasa-std-7009

O'Donnell TMP et al (2013) Revision of minimal resection resurfacing unicondylar knee arthroplasty to total knee arthroplasty. J Arthroplasty 28:33–39

Palumbo BT et al (2011) Initial experience of the Journey-Deuce bicompartmental knee prosthesis. J Arthroplasty 26(6):40–45
Patil S, Colwell CW Jr, Ezzet KA et al (2005) Can normal knee kinematics be restored with unicompartmental knee replacement? J Bone Joint Surg Am 87(2):332
Sarmah SS et al (2012) Annotation: knee the radiological assessment of total and unicompartmental knee replacements. J Bone Joint Surg Br 94:1321–1329
V&V 40 verification and validation in computational modeling of medical devices. https://cstools.asme.org/csconnect/CommitteePages.cfm?Committee=100108782

Effect of Natural Honey Treatment and External Stretching on Kinematics of Cell Migration During Gap Closure

Y. Berkovitch, S. Toume, A. Gefen and Daphne Weihs

Abstract Wound healing is a complex natural response to tissue injury intended to restore the integrity and function of the tissue at the wounded site. The wound healing process occurs naturally under physical/mechanical forces, e.g., local stretching, and in many cases, benefits from application of applied remedies. Identification of potentially effective compounds requires specialized approaches to determine their effects on different cell types. However, working in vivo has quantification limits and ethical concerns arise. This is especially true for natural origin compounds, such as plant extracts, honey, and larvae, which have a complex composition and determining their clinical efficacy is challenging. Thus, in vitro gap closure assays can be used to evaluate the effects of natural treatments and external strains on migration associated with gap closure. Here, we evaluate the changes in cell migration during gap closure in vitro, following treatment with honey and/or under applied external strains. We generated monolayers of NIH3T3 mouse fibroblasts and induced a small gap (wound) at their center. Using custom image processing modules, we measured the kinematics of the gap closure, focusing on times of initiation and rates of migration. We evaluated the effects of different concentrations of a well-known natural wound-care agent, i.e., honey. Honey has been used in skin wound care for many years due to its anti-inflammatory, antimicrobial, and cell-stimulating properties. Using our assay we are able to show the direct effects of honey on the migratory capabilities of the cells and the kinematics of gap closure. In addition, we evaluate the effects of externally applied stretching, with and without the addition of natural remedies, on the gap closure process. For this, we grow the cell monolayer on an elastic, stretchable membrane, which is then stretched in varying levels using a 3-D printed

Y. Berkovitch · S. Toume · D. Weihs (✉)
Faculty of Biomedical Engineering, Technion - Israel Institute of Technology, 3200003 Haifa, Israel
e-mail: daphnew@technion.ac.il

A. Gefen
Department of Biomedical Engineering, Faculty of Engineering, Tel Aviv University, 6997801 Tel Aviv, Israel
e-mail: gefen@eng.tau.ac.il

A. Gefen and D. Weihs (eds.), *Computer Methods in Biomechanics and Biomedical Engineering*, Lecture Notes in Bioengineering,
DOI 10.1007/978-3-319-59764-5_9

cell stretching apparatus. We evaluated the effects on gap closure kinematics of different levels of externally applied strains, showing that low (3%) strains may accelerate gap closure, while higher (6%) strain may not be as efficient, relative to unstretched control. Thus, combining external deformation with various treatments can enhance the rate of migration and thus shorten the time required for wound healing.

Introduction

Wound healing is a complex natural response to tissue injury intended to restore the integrity and function of the tissue at the wounded site. Although there are many types of wounds they undergo four, similar major healing stages: hemostasis, inflammation, proliferation, and remodeling. For proper recovery from damage in many cases, wounded tissue requires application of remedies developed or adapted for specific tissue, physical/mechanical conditions of the tissue, and supporting all healing stages. The cellular events that lead to wound healing occur concurretly, complicating isolation of affecting factors and demanding extensive research. Identification of potentially effective compounds requires specialized approaches to determine their effects on different cell types (Gefen and Weihs 2016). In vivo investigations have quantification limits and ethical concerns. This is especially true for natural origin compounds, such as plant extracts, honey, and larvae, which have a complex composition and determining their clinical efficacy is challenging (Topman et al. 2013). Thus, in vitro gap closure assays can be used to evaluate the effects of natural treatments and external strains on migration associated with gap closure.

Here, we evaluate the changes in cell migration during gap closure in vitro, following treatment with honey; or under applied external strains. Honey has been used in skin wound care for many centuries due to its anti-inflammatory, antimicrobial, and cell-stimulating properties. Using our assay, we are able to show the direct effects of honey on the migratory capabilities of the cells and the kinematics of gap closure. In addition, we evaluated the effects of externally applied stretching on the gap closure process, showing that low (3%) strains may accelerate gap closure, whereas higher (6%) strain may not be as effective, relative to unstretched control.

Materials and Methods

Cell Stretching Apparatus

We used a custom three-dimensionally printed stretching apparatus (Toume et al. 2016), based on the concepts, structure, and calibration of a previously machined version (Shoham et al. 2012). Briefly, we lower a 6-well culture plate with elastic

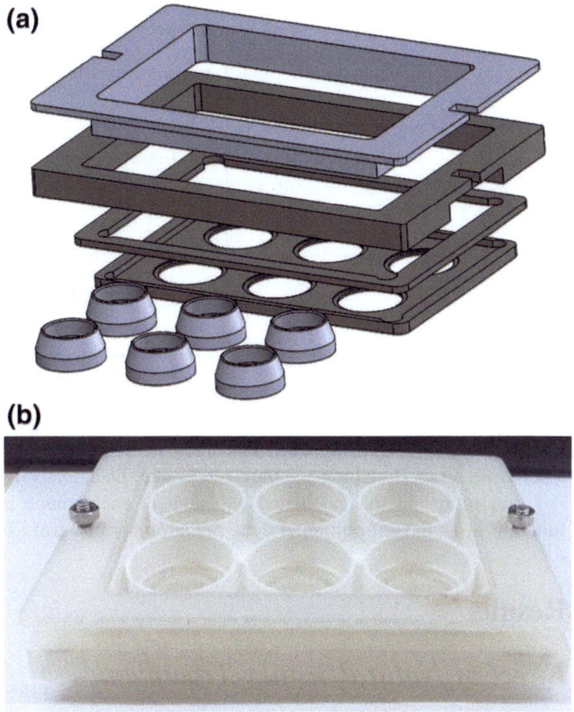

Fig. 1 Cell stretching device **a** schematic design of the apparatus parts, see Toume et al. (2016) for detailed description **b** The dynamic stretching device

substrata at its bottom on top of round inserts, both placed within two plates (Fig. 1). By controlling the distance between the top and bottom plates (using attached screws), we define the radial stretching applied to the substrata and the NIH3T3 fibroblasts (CRL-1658, ATCC) cells adhered to it; stretching is homogeneous.

Analysis of Gap Closure

We utilized a custom algorithm in MATLAB 2012b (MathWorks Inc., Nattick, MA) to automatically analyze the time-progression images of the gap area closure and quantify cell migration and gap closure progression (Fig. 2) (Topman et al. 2011, 2012a, b). The time-dependent area was fit to a Richard's function (1959), which is an asymmetric sigmoid, by minimizing the mean squared error. Using the fitted curve, we calculated the times for onset and end of mass cell migration, respectively, the times where 10% and 90% of the initial gap are closed, as well as the maximum and the average migration rates (Topman et al. 2012b).

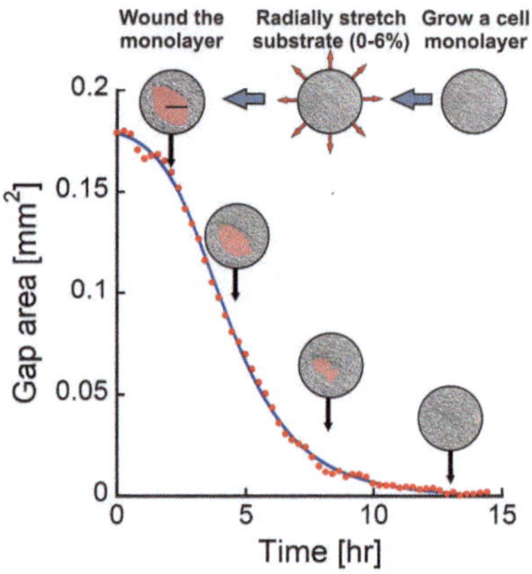

Fig. 2 Experimental design and data analysis. Schematic representation of stretching experiment, where instead of stretching we may add honey. Data analysis graph of gap area versus time. The *red dots* represent the calculated gap area in each image taken. The *blue line* is a Richard's function fit to the experimental data. *Scale bar* is 200 μm (Toume et al. 2017)

Results

Effect of Natural Honey Treatment

NIH3T3 mouse fibroblasts were allowed to grow until a cell monolayer formed. A gap was induced by crushing, using a truncated fiber-optic to tap the cell monolayer and then medium was replaced with medium containing desired concentrations of honey. The time-dependent gap closure by NIH3T3 migrating cells was monitored by time lapse microscopy and analyzed using custom image processing modules in MATLAB 2012b to detect the time-dependent denuded areas during the gap closure. Kinematic parameters of cell migration were calculated by fitting a Richards' function (Shoham et al. 2012) (see Fig. 2) focusing on times of initiation and rates of migration. We revealed that the cell migration rate with 0.02% (w/v) honey supplement was indistinguishable from control, yet a reduction when 0.2% (w/v) honey was added (not shown).

Effect of Different Levels of Externally Applied Deformations

The NIH3T3 cells were cultured on 6-well plate with an elastic, transparent, and collagen-coated bottom membrane till monolayer formation. The plate was mounted into custom made 3-D printed stretching device (Toume et al. 2016) (see Fig. 1). By lowering the plate on the cylindrical fixtures, different levels of radial stretching strains were induced (0, 3 or 6%). Immediately following stretching,

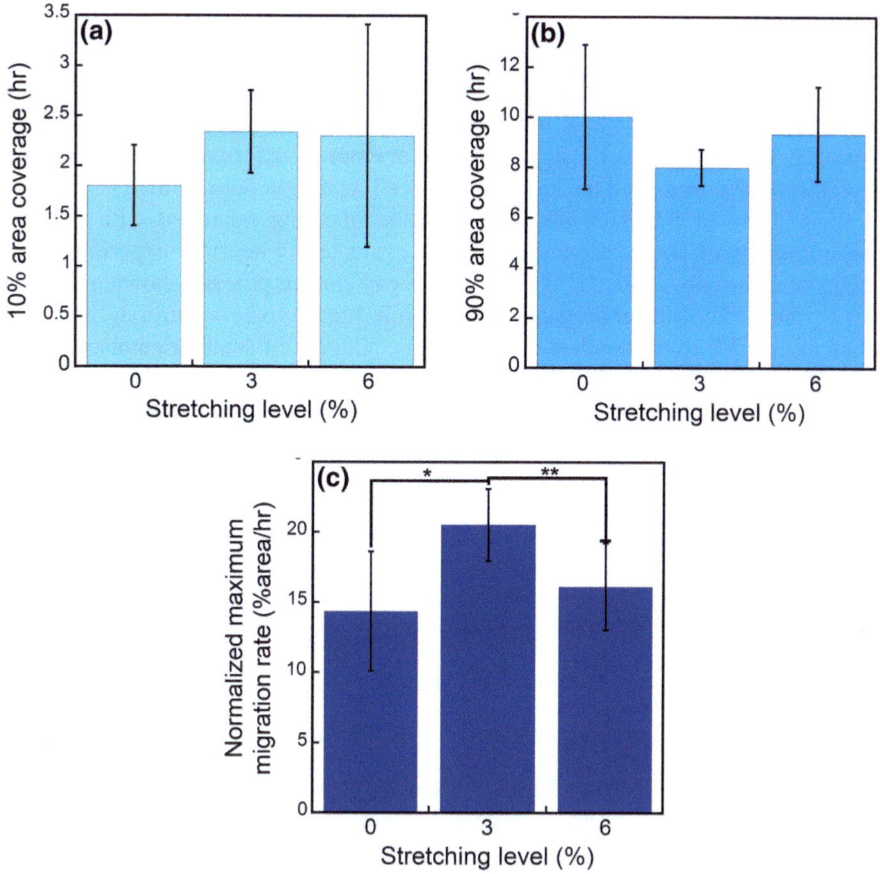

Fig. 3 Kinematics of the gap closure process, with injury being applied immediately prior to stretching. A total of 8 gap closure experiments are averaged for each stretching level, and the displayed *error bars* are standard deviations **a** The time for 10% coverage of the gap, indicating the onset of the *en mass* cell migration, as a function of stretching level **b** The time for 90% coverage of the gap, indicating the end of the *en mass* cell migration, as a function of the stretching level **c** The maximum migration rate for each stretching level normalized in respect to the initial wound area. Significant differences in the mean, maximum migration rates were observed between 0 and 3% ($p < 0.01$) and between 3 and 6% ($p < 0.05$) stretching levels (Toume et al. 2017)

the crushing gap was induced and kinematics were monitored and calculated (Toume et al. 2017). We observed a significant increase in normalized migration rates and reduction of gap closure time under 3% stretching, relative to unstretched controls, and also reduction of the effect under 6% stretch (Fig. 3).

Discussion

Wound healing occurs naturally under physical/mechanical forces, e.g., local stretching, and in many cases, benefits from applied remedies. Here, we utilize an approach to evaluate effectiveness of honey or external deformation in accelerating gap closure. We measured the kinematics of cell migration during small gap closure in monolayers of NIH3T3 mouse fibroblasts following treatment with different concentrations of honey. Concurrently, we evaluated in separate experiments the effects of externally applied stretching on the gap closure process, showing that low (3%) strains may accelerate gap closure, while higher (6%) strain may not be as efficient, relative to unstretched control. Thus, addition of small amounts of honey or application of low level radial stretching can both accelerate migration and gap closure.

Acknowledgments The work was partially supported by the Israeli Ministry of Science and Technology 3rd Age Program (3-12804).

References

Gefen A, Weihs D (2016) Mechanical cytoprotection: a review of cytoskeleton-protection approaches for cells. J Biomech 49:9–1321. doi:10.1016/j.jbiomech.2015.10.030

Richards FJ (1959) A flexible growth function for empirical use. J Exp Bot 10:290–301. doi:10.1093/jxb/10.2.290

Shoham N, Gottlieb R, Sharabani-Yosef O, Zaretsky U, Benayahu D, Gefen A (2012) Static mechanical stretching accelerates lipid production in 3T3-L1 adipocytes by activating the MEK signaling pathway. Am J Physiol Physiol 302:C429–C441

Topman G, Sharabani-Yosef O, Gefen A (2011) A method for quick, low-cost automated confluency measurements. Microsc Microanal 17:22–915. doi:10.1017/S1431927611012153

Topman G, Sharabani-Yosef O, Gefen A (2012a) A standardized objective method for continuously measuring the kinematics of cultures covering a mechanically damaged site. Med Eng Phys 34:32–225. doi:10.1016/j.medengphy.2011.07.014

Topman G, Lin FH, Gefen A (2012b) The influence of ischemic factors on the migration rates of cell types involved in cutaneous and subcutaneous pressure ulcers. Ann Biomed Eng 40:39–1929. doi:10.1007/s10439-012-0545-0

Topman G, Lin F-H, Gefen A (2013) The natural medications for wound healing—curcumin, aloe-vera and ginger—do not induce a significant effect on the migration kinematics of cultured fibroblasts. J Biomech 46:4–170. doi:10.1016/j.jbiomech.2012.09.015

Toume S, Gefen A, Weihs D (2016) Printable low-cost, sustained and dynamic cell stretching apparatus. J Biomech 49:9–1336. doi:10.1016/j.jbiomech.2016.03.026

Toume S, Gefen A, Weihs D (2017) Low-level stretching accelerates cell migration into a gap. Int Wound J 14(4):698–703. doi:10.1111/iwj.12679

3D Vessel Extraction in the Rat Brain from Ultrasensitive Doppler Images

E. Cohen, T. Deffieux, C. Demené, L.D. Cohen and M. Tanter

Abstract Ultrasensitive Doppler is a recent medical imaging technique enabling high sensitive acquisition of blood flows which can detect small vascular features without contrast agents. Applied to cerebral tomographic imaging of rodents, this method produces very fine vascular 3D maps of the brain at high spatial resolution of 100 μm. These vascular networks contain characteristic tubular structures that could be used as landmarks to localize the position of the ultrasonic probe and take advantage of the easy-to-use property of ultrasound devices. In this study, we propose a computational method that performs 3D extraction of vascular paths and estimates effective diameters of vessels, from ultrasensitive Doppler 3D reconstructed images of the rat brain. The method is based on the fast marching algorithm to extract curves minimizing length according to a relevant metric.

E. Cohen (✉) · T. Deffieux · C. Demené · M. Tanter
CNRS, UMR 7587, INSERM U979, ESPCI ParisTech, Institut Langevin,
PSL Research University, 75005 Paris, France
e-mail: emmanuel.cohen@espci.fr; emmanuel.cohen@dauphine.eu

T. Deffieux
e-mail: thomas.deffieux@espci.fr

C. Demené
e-mail: charlie.demene@espci.fr

M. Tanter
e-mail: mickael.tanter@espci.fr

E. Cohen · L.D. Cohen
CNRS, UMR 7534, CEREMADE, University Paris Dauphine,
PSL Research University, 75016 Paris, France
e-mail: cohen@ceremade.dauphine.fr

A. Gefen and D. Weihs (eds.), *Computer Methods in Biomechanics and Biomedical Engineering*, Lecture Notes in Bioengineering,
DOI 10.1007/978-3-319-59764-5_10

Introduction

Context

Medical ultrasound imaging has become a major clinical technique achieving an anatomical imaging of high quality, usually associated to a Doppler examination for blood flows observation and quantification. Ultrasound devices are also very convenient tools thanks to their portability, real-time working, and low cost. Today, the observation of very fast variations in the human body like mechanical waves propagation (Gennisson et al. 2013) or fast blood flows (Bercoff et al. 2011) is possible by ultrafast ultrasound imaging. Instead of using line-per-line focusing of ultrasonic beams like in standard ultrasound imaging, ultrafast imaging uses ultrasonic plane-wave transmissions associated to the power of graphical processing unit based platforms, in order to accelerate typical frame rates to more than 1000 frames per second (Tanter and Fink 2014). Ultrasensitive Doppler is one of these new ultrasensitive techniques which allows high sensitive acquisition of small vascular features without contrast agent.

Ultrasensitive Doppler imaging produces very fine 3D vascular maps of the rodent brain with high spatial resolution. The acquisition is realized in vivo thanks to a simple mechanical system described in Demené et al. (2016). A mechanical scanning process acquires successive 2D sections along the rodent brain surface for several different orientations. Using a tomographic approach, a postprocessing treatment of the data reconstructs a real 3D volume of the cerebral vascular network.

Motivations

The cerebral vascular network may serve as landmark in the brain for many purposes: real-time neuronavigation, neurosurgery monitoring, brain tumor monitoring, etc. For instance, in Cohen et al. (2016), we have proposed such a system to neuronavigate the rodent brain in real time. We successfully registered different ultrasensitive Doppler 3D scans using the cerebral vascular print as a matching feature. The algorithm used compares 2D brain sections by correlation, demonstrating the amount of positioning information contained by the ultrasensitive Doppler signal. Besides, this new real-time neuronavigation system based on ultrasounds is of great interest given the easy-to-use properties of ultrasound devices: portability, real time, low cost, etc.

Consequently, we envisage to extract some geometric features characterizing the cerebral vascular network from ultrasensitive Doppler 3D images. Indeed, a finer description of the vascular print will lead to a better understanding and analysis of the vasculature; in addition, it will serve as a more accurate matching feature in all registration tasks aiming to find invariant cerebral vascular structures. In the present

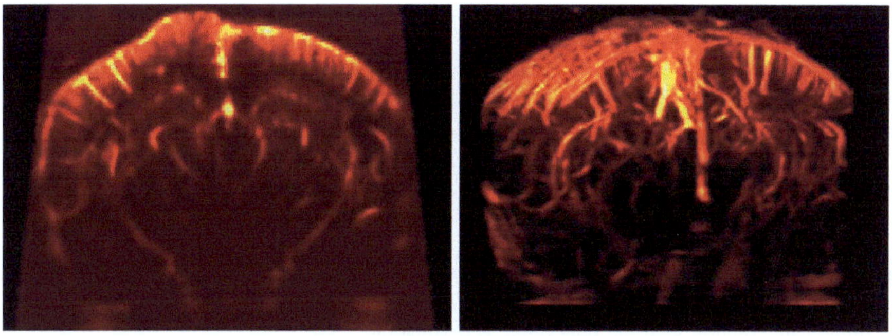

Fig. 1 Ultrasensitive Doppler imaging of the rat brain. *Left* 2D ultrasensitive Doppler coronal plane. *Right* 3D tomographic reconstruction (Demené et al. 2016)

study, we propose a computational method that performs 3D extraction of vascular paths from tomography based 3D reconstruction of the rat brain by ultrafast ultrasound imaging (Demené et al. 2016). In order to distinguish between small and large vessels, the method also enables the estimation of vessel diameters (Fig. 1).

Materials and Methods

Material

In vivo experiments were performed on anesthetized rats using ultrasensitive Doppler. Ultrafast ultrasonic imaging enables fast acquisition of brain sections at 100 μm × 100 μm resolution in the image plane. A 15 MHz motorized probe acquires 400 μm-thick brain sections with 200 μm spacing. A typical 3D scan of the total width of the rat brain along one specific direction contains around 65 sections. A tomographic reconstruction can be achieved from several scan acquisitions along 18 different orientations, to obtain 3D high spatial resolution volume of around 200 × 300 × 300 pixels of size 100 μm × 100 μm × 100 μm A complete description of the experimental setup can be found in Demené et al. (2016).

Minimal Path Extraction

The problem of extracting the minimal path between two points $\mathbf{x_0}, \mathbf{x_1} \in \Omega \subset \mathbb{R}^d$ $(d = 2, 3)$ in an image consists in finding the curve joining them and following the shortest possible path. The definition of a shortest path is generally associated to the Euclidean metric (or distance) $d_e(\mathbf{x_0}, \mathbf{x_1}) = \langle \mathbf{x_0}, \mathbf{x_1} \rangle^{1/2}$ leading to the segment

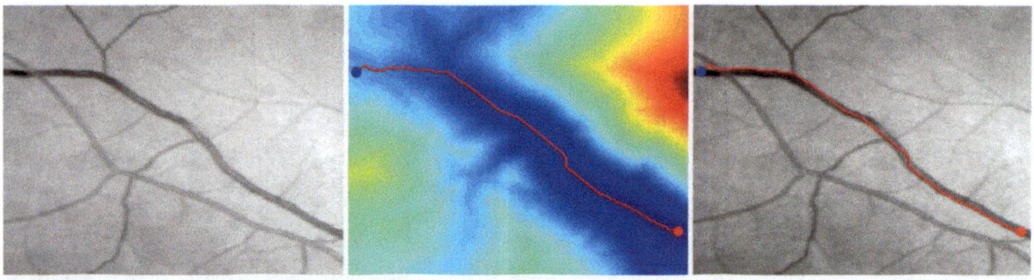

Fig. 2 Illustration of the minimal path extraction technique used on a 2D retina image. From *left* to *right*: the original image; the geodesic distance map corresponding to the *red start point*; the minimal path solution with the *blue point* as destination

joining $\mathbf{x_0}$ and $\mathbf{x_1}$ for the minimal path. However, in an image, a segment does not generally suite to represent the path between two points (see for instance the extraction of a retina vessel on Fig. 2).

Therefore, one can introduce a Riemannian metric that locally depends on the pixel intensity in order to make shortest paths depend on the most salient variations of the image, as it is well explained in Peyré et al. (2010). Let $\gamma : [0,1] \to \mathbb{R}^d$ be a smooth curve joining $\gamma(0) = \mathbf{x_0}$ to $\gamma(1) = \mathbf{x_1}$, a Riemannian metric or geodesic distance can be defined as

$$d(\mathbf{x_0}, \mathbf{x_1}) = \min_{\gamma} L(\gamma),$$

where $L(\gamma)$ is the Riemannian curve length of γ, and is given by

$$L(\gamma) = \int_0^1 \langle \gamma'(t), W(\gamma(t))\gamma'(t) \rangle^{1/2} \mathrm{d}t.$$

As expected, the metric is now a function of a saliency map $W : \mathbb{R}^d \to \mathbb{R}$ called potential that should be chosen low in the image areas to be extracted. In general, W is a symmetric positive tensor field of $\mathbb{R}^{d \times d}$ but we restrain our study to the isotropic case. In fact, to extract vessels, we will build W from the intensity of the blood flow which is a scalar function of the 3D space.

From then on, the minimal path extraction problem is to solve the following minimization problem

$$\gamma^* = \arg\min_{\gamma} L(\gamma). \tag{1}$$

Given any set S of start points, one can define the geodesic distance map by

$$U_S(\mathbf{x}) = \min_{\mathbf{x_0} \in S} d(\mathbf{x_0}, \mathbf{x}).$$

As described in Cohen and Kimmel (1997) and Peyré et al. (2010), if we manage to compute U_S, the solution of the problem (1) satisfies the following gradient descent

$$\frac{\mathrm{d}\gamma^*(t)}{\mathrm{d}t} = -\frac{\nabla U_S(\gamma^*(t))}{\|\nabla U_S(\gamma^*(t))\|}. \tag{2}$$

The computation of U_S is more complicated. From results on the viscosity solution of the Hamilton–Jacobi equation (Crandall and Lions 1983; Peyré et al. 2010), it can be proved, under assumptions of compactness and continuity respectively for S and W, that U_S is the unique viscosity solution of the Eikonal equation

$$\begin{cases} \|\nabla U_S(\mathbf{x})\| = W(\mathbf{x}), & \forall \mathbf{x} \in \Omega, \\ U_S(\mathbf{x}) = 0, & \forall \mathbf{x} \in S. \end{cases} \tag{3}$$

Some geometric interpretations can also be found in Cohen and Kimmel (1997) and serve as elements of proof for better understanding.

To solve numerically the Eikonal equation, we use the fast marching method (FMM) introduced by Sethian (1999) and we follow the numerical scheme presented in Deschamps and Cohen (2001) for 3D images ($d = 3$). The FMM is a front propagation approach computing iteratively the values of U_S in increasing order. Then, we extract γ^* by using a simple discrete gradient descent to solve (2) avoiding angular error accumulation (Cohen and Kimmel 1997): from the destination point $\mathbf{x_1}$, the back propagation follows the direction of the lowest U_S value of a regular neighbor grid around the current iteration point, until the first start point $\mathbf{x_0}$ is reached, i.e., as soon as $U_S(\mathbf{x}) = 0$.

Illustrations of the method just described in 2D and 3D are shown in Figs. 2 and 3.

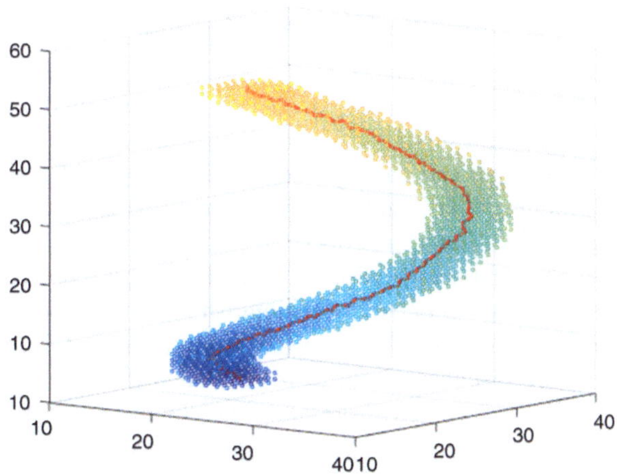

Fig. 3 Illustration of the minimal path extraction technique used on a 3D synthetic helical structure. The start and destination points are respectively chosen at the *bottom* and *top* of the helice. The *color map* follows the values of the geodesic distance map

Iterative Keypoint Detection

In the case of vessel extraction and particularly when a large amount of vessels constitutes a complex vascular network, it is convenient to extract the desired vessels from a unique start point. Indeed, because of bifurcations of vessels, it would require many different unknown pairs of start and destination points of the minimal path extraction method described just before.

Therefore, we use the Minimal Path method With Keypoint Detection (MPWKD) introduced by Benmansour and Cohen (2009). From a single start point $\mathbf{x_0}$ and given a parameter λ, intermediate points, called keypoints, are successively detected along the curve of interest with a spacing between them almost constant to λ. Another advantage of this method is the influence of λ on the path when the potential is too noisy or not enough contrasted; in this case, small values of λ can prevent minimal path solution from wrong shortcuts to the start point (Benmansour and Cohen 2009).

Like FMM, the MPWKD algorithm is based on minimal path extraction technique described in Section "Minimal Path Extraction". The geodesic distance map $U_{\mathbf{x_0}}$ is first computed until the front reaches a keypoint $\mathbf{p_0}$ with an Euclidean distance higher than λ. Then, the front continues to propagate considering $\mathbf{p_0}$ as a new start point, i.e., $S = \{\mathbf{x_0}, \mathbf{p_0}\}$ and $U_S(\mathbf{p_0}) = 0$, until a new keypoint $\mathbf{p_1}$ again with an Euclidean distance higher than λ. The algorithm iterates this procedure until the approximated total Euclidean length of the path is higher than a second parameter L to obtain a quasi-equidistant set of keypoints $\mathbf{p_i}$. The Euclidean distance map is computed just as the geodesic by solving (3) with $W(\mathbf{x}) = 1$ for all $\mathbf{x} \in \Omega$. The extraction of the total minimal path is performed at each iteration after which a new keypoint is detected by the same simple gradient descent on U_S as in Section "Minimal Path Extraction". The MPWKD is also taking into account the number of adjacent Voronoi regions in order to extract the correct path.

Vessel Effective Diameter Estimation

From the previous analysis, the obtained minimal path extraction can be viewed as a graph representation of the vascular network. The keypoints would be the nodes of the graph and an edge would be defined by the path connecting two successive keypoints. This graph description may be particularly useful to compare different cerebral vascular networks.

As a first step leading to a suitable graph representation of the cerebral vascular network, we propose to distinguish vessels according to their apparent diameters. Indeed, the size of the vessels is a crucial feature useful to characterize different kind of vascular structures in relation with some physiological parameters such as blood flow velocity, pressure, etc. In the context of ultrafast Doppler imaging, due to the physics of ultrasound devices, the observed size of vessels is not

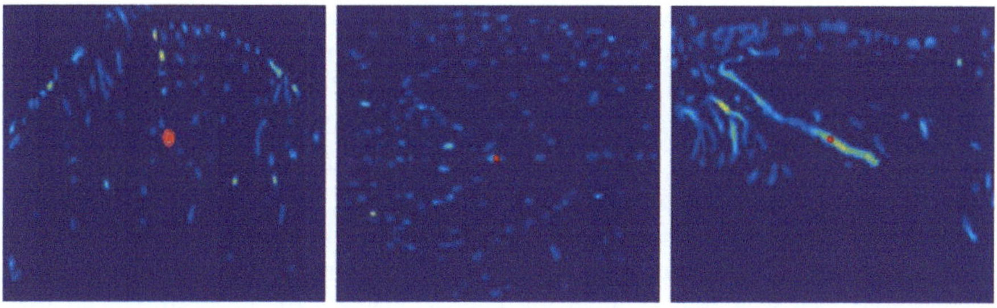

Fig. 4 Vessel apparent diameter estimation method. From *left* to *right*: coronal, horizontal, and sagittal planes that locally intersect the minimal path at the *red point*. In this example, the smallest fitted circle is on the coronal plane

rigorously their real size. That is why we will deal here with apparent diameters of vessels.

The previous extracted minimal paths are supposed to be localized inside the vessels, even if they are not a priori centered on their centerlines (Peyré et al. 2010). Thus, at each point of the extracted minimal paths, one can consider the orthogonal plane to the vessel and passing through this point. In this plane, if we model a vessel by a curvilinear and tubular structure, the point of interest should be located inside a group of pixels fitting an ellipse. The vessel diameter is then estimated by the characteristic size of the latter ellipse.

Computing the orthogonal plane requires to know precisely the local orientation of the vessel. A simple faster method consists in considering the best approximation of the orthogonal plane among the local coronal, sagittal, and horizontal sections. This is done by fitting an ellipse or basically a circle in each of those three planes and selecting the smallest one as the correct circle estimate (see Fig. 4); in fact, the two other bigger estimates fit actually noncircular structures revealing the non-orthogonality to the vessel. Finally, the center of the fitted circle allows to recenter the point of interest and thus the minimal path at the center of the vessel.

Results and Discussion

In order to show the efficacy of the proposed method, we first apply it on a 2D coronal plane of the rat brain (Fig. 6), obtained by averaging 30 successive coronal sections of the 3D ultrafast Doppler tomographic reconstruction presented in Section "Material". A Doppler image shows the intensity of the blood flow which is maximal at the center of the vessels. Therefore, the metric should be simply chosen as an inversely proportional function to the image. A first simple and successful model corresponds to $W = 1/I$, where I is the 2D or 3D ultrafast Doppler image. Before launching the keypoints detection algorithm (see Section "Iterative Keypoint Detection"), one must take into account the image background the presence with many pixels of very low intensity that do not correspond to vessels. Otherwise, the MPWKD finds many outlier keypoints. The left image of Fig. 5 illustrates this over-detection result. Then, the middle image shows the MPWKD performance after a thresholding of the data: a large part of the vascular network is well extracted revealing the importance of the preprocessing segmentation step of the image. The right image shows how to get better performance by performing before the thresholding step a histogram equalization based on Zuiderveld (1994) in order to increase the contrast and then enhance vascular structures. This time, almost all the 2D vascular tree is extracted even the deepest vessels of the brain. In all these simulations, the start point from which the MPWKD starts the keypoint detection is red colored and chosen in an image area of highest intensity like the vertical central vessel located on the symmetry axis of the brain. One can appreciate the enhancement of the vascular structures of the image on Fig. 6. The two last images of Fig. 6 show the Euclidean distance map updated at each keypoint detection and the approximated total Euclidean length.

The proposed method was then tested in 3D on a tomographic reconstruction of the rat brain (see Section "Material"). In real 3D data, the observed vascular structures are much more consistent with the real vasculature, which was not the case in 2D average of several successive planes where wrong superimposition and then bifurcations of vessels could appear. That is why the MPWKD produces a very nice 3D vascular network extraction even without thresholding of the data (Fig. 7).

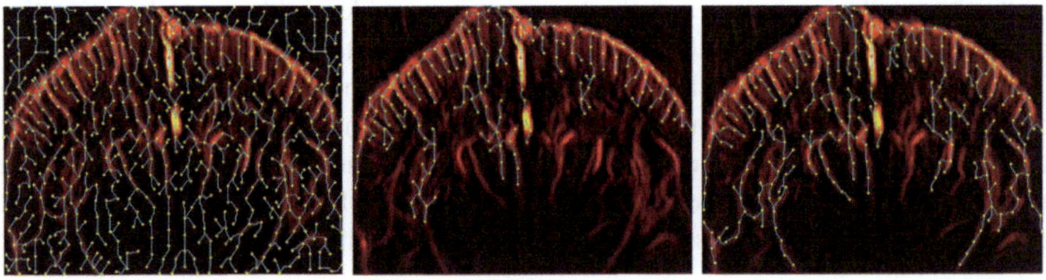

Fig. 5 Minimal path extraction with keypoint detection in 2D. From *left* to *right*, the method is applied to: original image; image after thresholding; image after thresholding and histogram equalization. Start points, keypoints, and minimal paths are respectively marked in *red*, *yellow*, and *cyan*

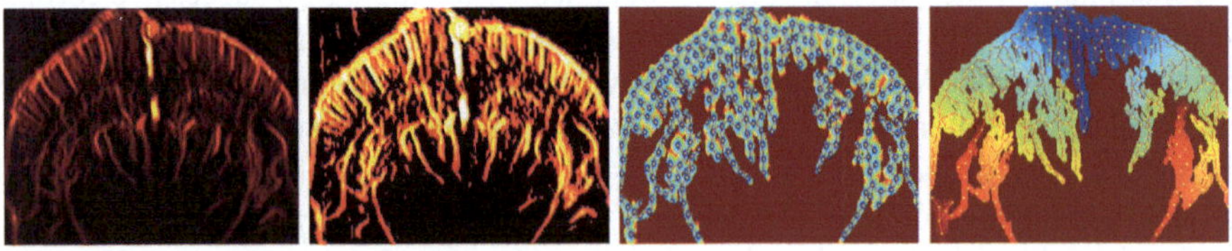

Fig. 6 From *left* to *right*: 2D ultrasensitive Doppler; thresholding and histogram equalization of the previous image; updated Euclidean distance map; approximated total Euclidean length map

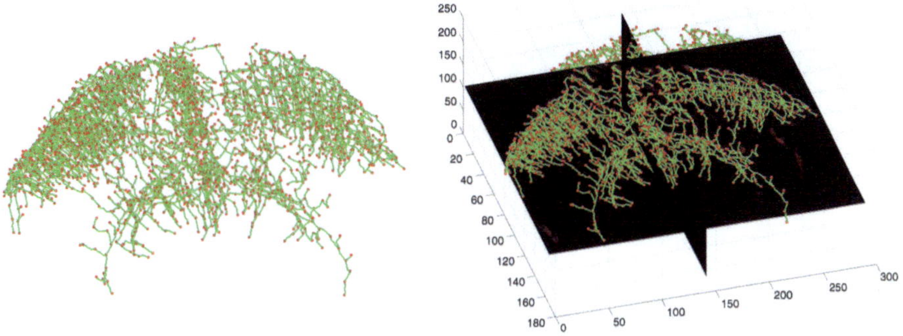

Fig. 7 Minimal path extraction with keypoint detection in 3D. Start points, keypoints, and minimal paths are respectively marked in *blue*, *red*, and *green*

Fig. 8 Vascular segmentation in 3D by the optimally oriented flux filter

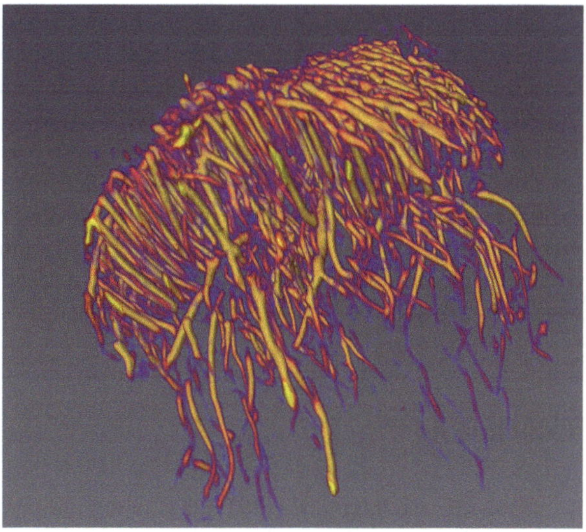

One can observe the profusion of extracted vessels in particular the high number of short vessels that makes difficult the physical interpretation and validation of the result. In order to refine the extraction, a preprocessing step to segment the more significant vascular structures can be applied. For instance, we used the optimally oriented flux (OOF) filter proposed by Law and Chung (2008) for curvilinear structure detection. Figure 8 illustrates the 3D segmentation of the data by OOF. Many noisy pixels are changed to zero and the surface of vessels is more accurately delimited. However, when the MPWKD is directly applied on the segmented volume, fewer keypoints are detected and one may repeat the keypoint detection algorithm from different start points to get the whole vascular tree.

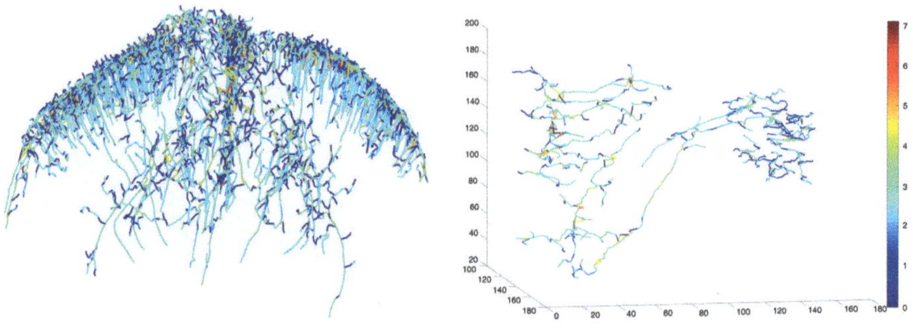

Fig. 9 Vessel apparent diameter estimation. *Left* vessel diameters of the whole extracted vascular network. *Right* vessel diameters of a partial vascular network computed by MPWKD after OOF filtering (the forebrain is on the *left* of the graph). The range of the color map values is from 0 (*blue*) to 7 (*red*) pixels

The OOF filter is particularly useful for the vessel apparent diameter estimation (see Section "Vessel Effective Diameter Estimation"). Indeed, in order to get consistent circle fitting like on Fig. 4, the vessel section containing the red point should be composed of almost exclusively pixels of interest, since the algorithm estimating the diameter only counts those pixels to define the area of the vessel section. If this area is not well delimited, then many outlier pixels will lead to a wrong estimated area. Thus, after delimiting vessels by OOF and using the vascular tree obtained by MPWKD on original data, we obtain a color map of the estimated vessel diameters as shown on Fig. 9. On can observe the profusion of small vessels that ultrafast ultrasound imaging enables to detect.

Conclusion

In this study, we have presented a first computational method capable of analyzing the vasculature of the rat brain from 3D ultrasensitive Doppler. The proposed method has the advantage to be really fast (tens of seconds), implementing the solution with C++. Future work should improve several issues: the accuracy of the minimal path extraction by considering anisotropic metrics and studying the influence of the distance parameter between the keypoints; the enhancement of vascular structures to improve the quality of the extracted vascular tree; and finally, the accuracy of the proposed vessel diameter estimation by considering local orthogonal planes to vessels.

References

Benmansour F, Cohen LD (2009) Fast object segmentation by growing minimal paths from a single point on *2d* or *3d* images. J Math Imaging Vis 33(2):209–221

Bercoff J, Montaldo G, Loupas T, Savery D, Meziere F, Fink M, Tanter M (2011) Ultrafast compound doppler imaging: providing full blood flow characterization. IEEE Trans Ultrason Ferroelectr Freq Control 58(1):134–147

Cohen LD, Kimmel R (1997) Global minimum for active contour models: a minimal path approach. Int J Comput Vis 24(1):57–78

Cohen E, Deffieux T, Tiran E, Demene C, Cohen L, Tanter M (2016) Ultrasensitive doppler based neuronavigation system for preclinical brain imaging applications. In: 2016 IEEE international ultrasonics symposium (IUS). IEEE, Berlin, pp 1–4

Crandall MG, Lions PL (1983) Viscosity solutions of Hamilton–Jacobi equations. Trans Am Math Soc 277(1):1–42

Demené C, Tiran E, Sieu LA, Bergel A, Gennisson JL, Pernot M, Deffieux T, Cohen I, Tanter M (2016) 4d microvascular imaging based on ultrafast doppler tomography. NeuroImage 127:472–483

Deschamps T, Cohen LD (2001) Fast extraction of minimal paths in 3d images and applications to virtual endoscopy. Med Image Anal 5(4):281–299

Gennisson JL, Deffieux T, Fink M, Tanter M (2013) Ultrasound elastography: principles and techniques. Diag Interv Imaging 94(5):487–495

Law MW, Chung AC (2008) Three dimensional curvilinear structure detection using optimally oriented flux. In: European conference on computer vision. Springer, Berlin, pp 368–382

Peyré G, Péchaud M, Keriven R, Cohen LD (2010) Geodesic methods in computer vision and graphics. Found Trends® Comput Graph Vis 5(3–4):197–397

Sethian JA (1999) Level set methods and fast marching methods: evolving interfaces in computational geometry, fluid mechanics, computer vision, and materials science, vol 3. Cambridge University Press, Cambridge

Tanter M, Fink M (2014) Ultrafast imaging in biomedical ultrasound. IEEE Trans Ultrason Ferroelectr Freq Control 61(1):102–119

Zuiderveld K (1994) Contrast limited adaptive histogram equalization. In: Graphics gems IV. Academic Press Professional Inc., San Diego, CA, pp 474–485. http://dl.acm.org/citation.cfm?id=180895.180940

References

Fluid–Structure Simulation of a Transcatheter Aortic Valve Implantation: Potential Application to Patient-Specific Cases

W. Wu, D. Pott, C. Chiastra, L. Petrini, G. Pennati, G. Dubini, U. Steinseifer, S. Sonntag, M. Kuetting and F. Migliavacca

Abstract Valve diseases are more and more treated with transcatheter aortic valves. This work is based on an experimental setup with the corresponding fluid–structure interaction model to show the feasibility of performing accurate simulations which is able to capture the main behavior of a transcatheter valve both from structural and fluid dynamic points of view. The application of this methodology to patient-specific cases is also illustrated.

Introduction

Transcatheter aortic valve (TAV) (Tchetche et al. 2014; van Gils et al. 2014) implantation is a mini-invasive procedure adopted in the treatment of valve diseases. The behavior of a TAV is influenced by the mechanics of the aortic root, the leaflets, the stent frame, and the fluid passing through the site where the valve is implanted. This work presents a fluid–structure interaction (FSI) model for the evaluation of Nitinol TAVs.

W. Wu · C. Chiastra · L. Petrini · G. Pennati · G. Dubini · F. Migliavacca (✉)
Laboratory of Biological Structure Mechanics, Chemistry Materials and Chemical Engineering 'Giulio Natta' Department, Politecnico Di Milano, Milan, Italy
e-mail: francesco.migliavacca@polimi.it

D. Pott · U. Steinseifer · S. Sonntag · M. Kuetting
Department of Cardiovascular Engineering, Institute of Applied Medical Engineering, University Hospital Aachen, Aachen, Germany

D. Pott · U. Steinseifer · S. Sonntag · M. Kuetting
RWTH Aachen University, Aachen, Germany

A. Gefen and D. Weihs (eds.), *Computer Methods in Biomechanics and Biomedical Engineering*, Lecture Notes in Bioengineering,
DOI 10.1007/978-3-319-59764-5_11

Materials and Methods

A TAV (Figs. 1, and 2), after it was used in an in vitro mock loop (Fig. 3), was virtually implanted in a reconstruct aortic root simulating in vivo conditions (Fig. 4). The Nitinol TAVs are equipped with polyurethane leaflets and the model was built according to their dimensions. All material properties of the valve were acquired through experimental tests. As boundary conditions, the typical ventricular and aortic pressure curves were applied to the inlet and outlet sections, respectively (Fig. 5). The leaflets and compartment were coupled to the domain during the simulation. The commercial explicit finite element solver LS-DYNA was used to run the FSI simulation.

Results

The comparison between in vitro mock loop valve kinematics and the FSI corresponding results were in good agreement (Fig. 6). The mean strain and strain amplitude of the valve predicted by FSI simulation considering the aortic root (Fig. 7), if compared to a case where the valve is implanted in a silicon cylindrical compartment for accelerated fatigue tests, showed higher strains potentially dangerous for the fatigue behavior.

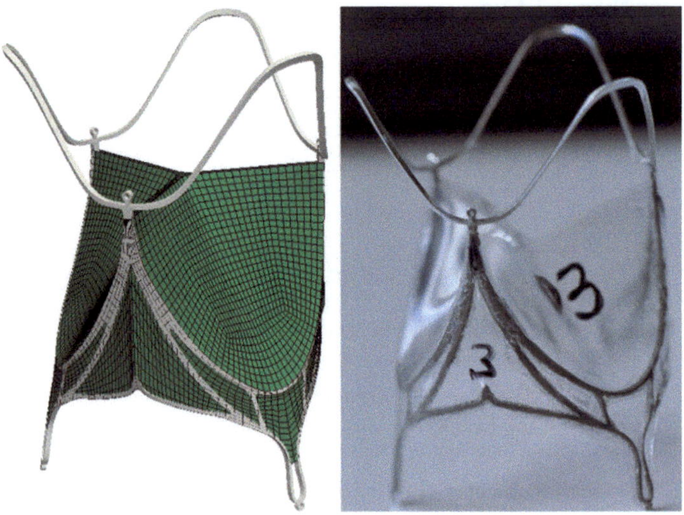

Fig. 1 Nitinol TAV (*right*) and corresponding finite element model (*left*)

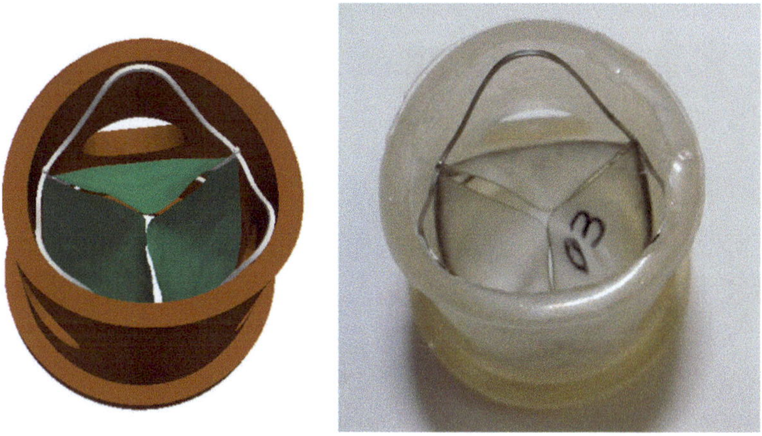

Fig. 2 TAV inserted in a silicon compartment (*right*) and corresponding finite element model (*left*)

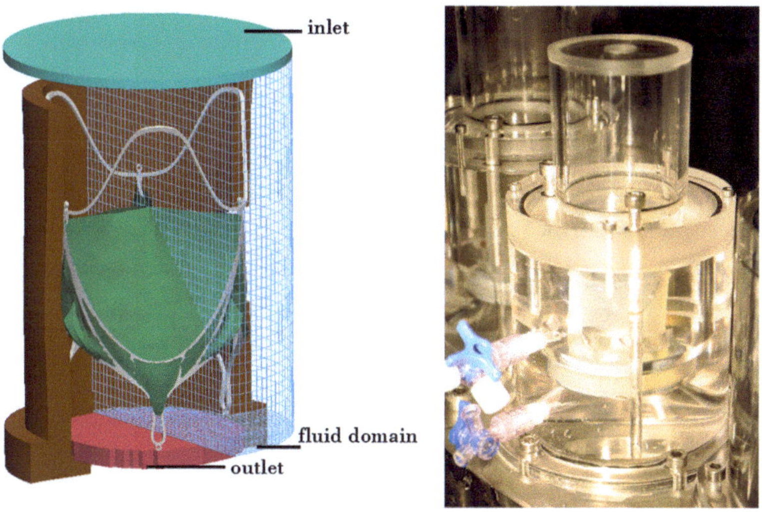

Fig. 3 In vitro mock loop (*right*) and corresponding fluid–structure interaction model (*left*)

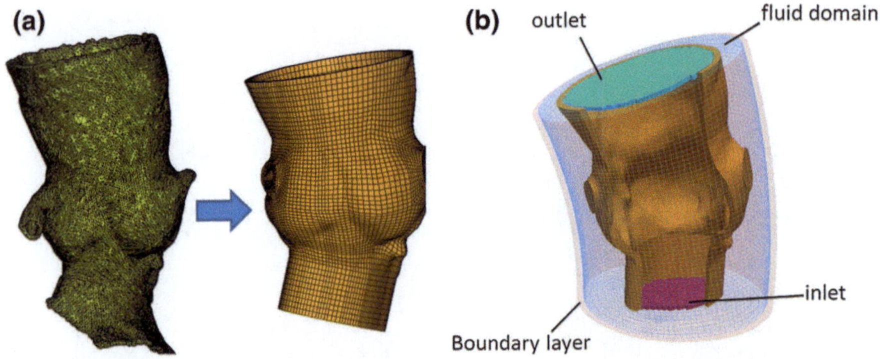

Fig. 4 **a** 3D aortic root anatomy of stereo lithography (STL) format was transformed to hexahedral solid elements. **b** Fluid–structure interaction model for the in vivo case

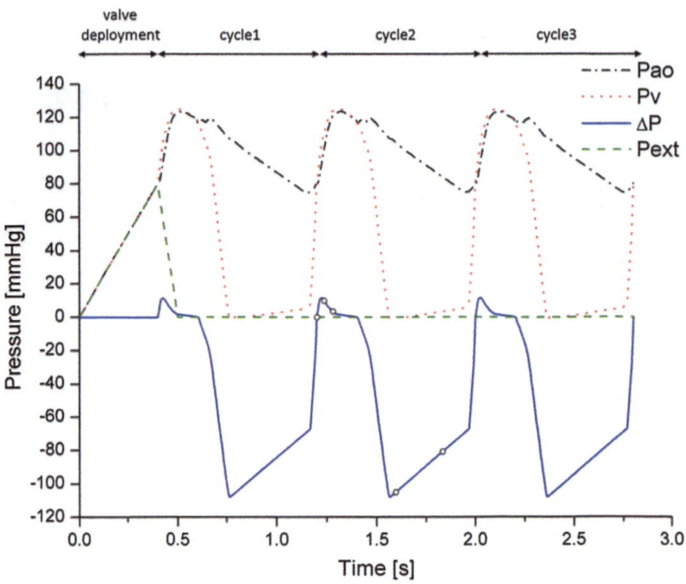

Fig. 5 Aortic (ao), ventricular (v), and extramural (ext) pressure tracings used in the patient-specific FSI simulation. The transvalvular gradient (ΔP) is also reported

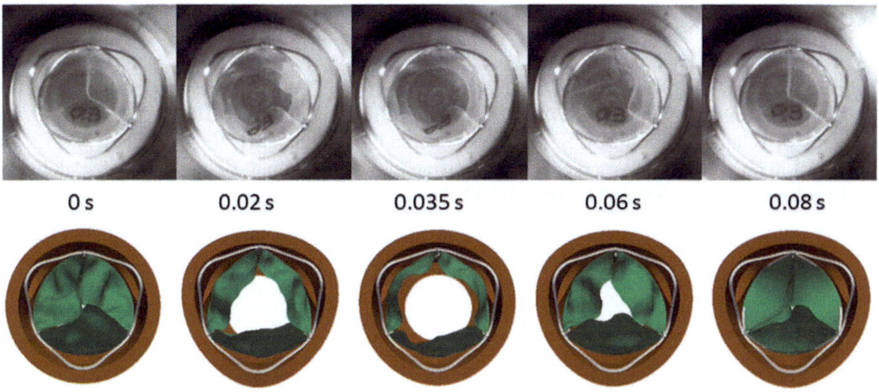

Fig. 6 Top views of the valve kinematics at five time instants for the simulation in vitro test (*top*) and the FSI (*bottom*)

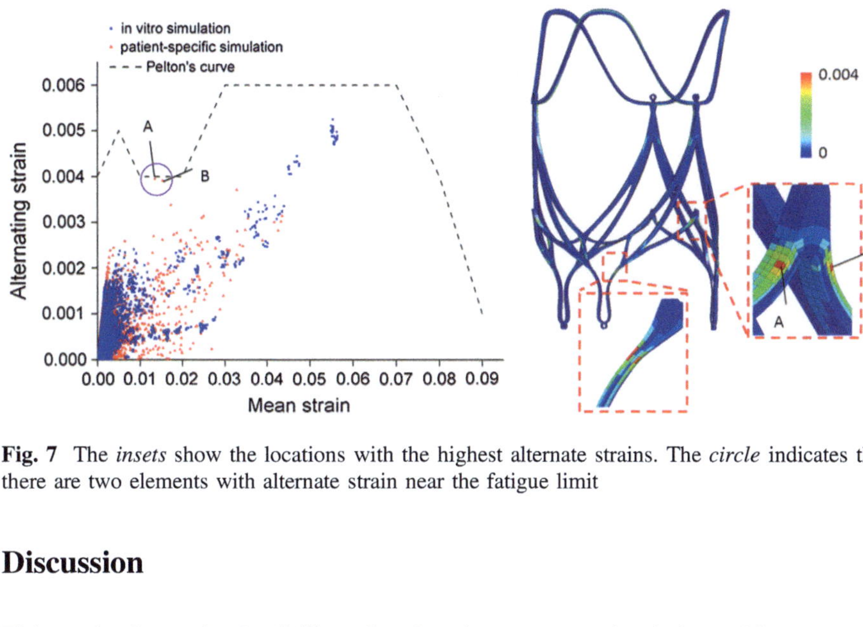

Fig. 7 The *insets* show the locations with the highest alternate strains. The *circle* indicates that there are two elements with alternate strain near the fatigue limit

Discussion

This study shows the feasibility of performing accurate simulations able to capture the main behavior of a TAV both from structural and fluid dynamic points of views. The application of this methodology to patient-specific cases enhances the capability to perform a correct surgical planning.

References

Tchetche D, Van Mieghem NM (2014) New-generation TAVI devices: description and specifications. EuroIntervention (2014) 10:U90–U100. EuroIntervention (2016) 12(Y):Y22–Y27

van Gils L, Tchetche D, Latib A, Sgroi C, Manoharan G, Möllmann H, Van Mieghem NM (2014) TAVI with current CE-marked devices: strategies for optimal sizing and valve delivery. EuroIntervention 10(Suppl U):U90–U100

Continuum-Scale Models for the Evolution of Hypertrophic Scars and Contractions After Burn Injuries

Fred Vermolen and Daniël Koppenol

Abstract We review several of our mathematical models that we constructed for the simulation of contractures and morpho-elastic scars that are typically associated with deep dermal (burn) injuries. The models are based on partial differential equations, which are solved by the use of finite-element methods. The models contain elements of non-isotropy, morpho-elasticity for the treatment of the mechanics of the skin. Furthermore, we take into account the balances of fibroblasts, myofibroblasts, collagen and a generic growth factor. Using the models, we are able to simulate permanent contractions using physically sound principles.

Introduction

Over the globe, about eleven million individuals are affected by burn injuries. In about six hundred thousand cases, the injury is so serious that the patient dies. After serious burns or incisions, hypertrophic scars may arise among patients, who need plastic surgery. Another side effect of burn injuries is the formation of contractures, which lead to permanent deformations and stresses in the skin of the patient. This causes a reduction of the mobility of the patient. This reduction is a consequence of the pulling forces that are exerted by the fibroblasts and myo-fibroblasts. These cells pull the surrounding tissue, by which the wound is contracted inwardly. This process is referred to as *contraction*. In the case that the deformations are small, the skin will recover mainly after the burn injury has healed. If, however, the deformations are large, which may happen in deep, serious burn injuries, the skin will no longer be able to recover. Then the patient suffers from a permanent deformation. If this deformation also deteriorates the mobility of the patient, then one speaks of a *contracture*.

F. Vermolen (✉) · D. Koppenol
Delft Institute of Applied Mathematics, Delft University of Technology,
2628 CD, Delft, The Netherlands
e-mail: F.J.Vermolen@tudelft.nl

A. Gefen and D. Weihs (eds.), *Computer Methods in Biomechanics and Biomedical Engineering*, Lecture Notes in Bioengineering,
DOI 10.1007/978-3-319-59764-5_12

During the past decades, mathematical modeling has progressed considerably in simulating wound healing, organ development, wound contracture, healing of bones and in the initiation and spread of cancer. This kind of models is applied on several scales, which give models that are based on entirely different principles, between stochastic processes and fully deterministic processes in terms of partial differential equations. The models on the smallest scales treat cells as individual entities, or even treat one cell (or even just several parts of it) only. Other models are based on the migration and deformation of cells. One may think, for instance, of cancer cells which migrate through small apertures in extracellular material and through blood vessels in order to spread (metastasize) over the body of the patient. At a somewhat larger scale, one can consider cell colonies in which each individual cell has a predefined geometry, and where it is able to divide, die, migrate or to differentiate to another phenotype. Here, one may distinguish between models in which cells can take whatever spatial position in a given domain, and these models where can only be located on discrete lattice points (such as cellular automata models, in which cellular Potts models form an important subclass). We refer to Vermolen (2016) for a review of such models. If one wants to take each cell into account when simulating a biomechanical mechanism like the formation of contractures, then one needs a very powerful computational environment with a huge amount of memory. In general, this is hard to achieve, and therefore, continuum-based models, which are based on cell densities, are currently being developed.

In this manuscript, we will concentrate on several mathematical models that are based on continuum-scale formalisms, which are applied to hypertrophic scars and contractures in relation to burn injuries. The current manuscript is based on the modeling studies in Koppenol et al. (2016a, b), Koppenol and Vermolen (2017) and therewith it should be seen as an advertisement for reading the aforementioned papers.

Materials and Methods

In this manuscript, we only present a summary of our modeling efforts in the context of burn injuries, where we consider a sequence of results for hypertrophic scars and contractures. To put the modeling work into its context, we briefly describe the biological stages a skin goes through after damage that was inflicted. The healing process starts with haemostasis. This is a process that stops bleeding, in which small blood vessels are constricted, where platelets converge and in which extracellular material is deposited. The platelets are responsible for the secretion of chemokines, which influence the behavior of immune cells, fibroblasts and endothelial cells. Subsequently, the immune cells enter the wound, and then the inflammatory phase commences. The immune cells clear up the contaminants and neutralize harmful pathogens. Shortly after this partial process, the proliferative phase sets in, where wound closure, wound contraction, angiogenesis are important

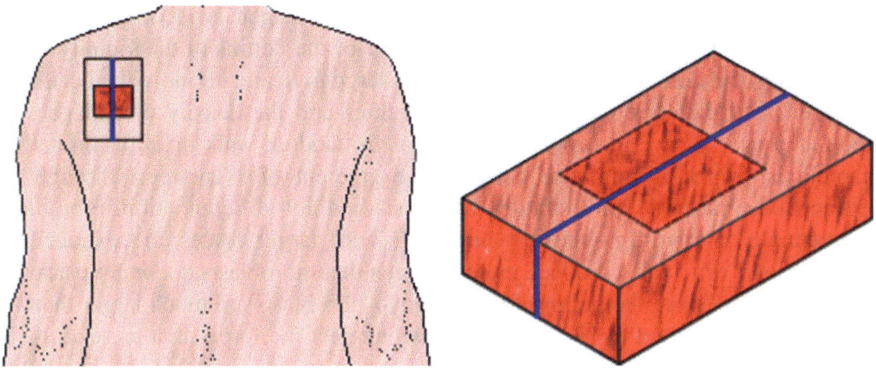

Fig. 1 Schematic of a burn injury on the shoulder (*left*), and a simplification of the geometry (*right*)

sub-processes. The epidermis is regenerated in the epithelialization process. Herewith, a protective layer over the wound is reconstructed. Furthermore, the dermis is reconstructed under the epidermis, where the dermis has a different texture than the epidermis and the texture of the dermis differs from an embryonic skin pattern. An important difference is the isotropy of the skin. Embryonic skin contains collagen molecules that are ordered more or less randomly, which allow an isotropic treatment of the skin. However, newly generated skin possesses an orientation that is linked to the migration pathways of fibroblasts. Herewith, the new extracellular material no longer contains an isotropic structure.

Finally, the remodeling phase takes place. During this phase, which in general lasts much longer than the other phases, the structure of the extracellular material changes. This leads to changes of the chemical composition and in the orientation of the collagen. Further, the number of fibroblasts and endothelial cells decrease during this phase.

First the tissue that is considered is simplified. We assume that the tissue that contains the damaged and undamaged part is represented correctly in a domain of computation. Such a domain of computation consists of the wound and a considerable portion of undamaged tissue, see Fig. 1, where we show a model for a burn injury near the shoulder of the patient.

Mathematical models are always simplifications of reality, since it is impossible to incorporate all the biological processes in detail. There exist several reasons for this, first, not all biological mechanisms regarding the formation of hypertrophic scars and contractures are known. The behavior of all cells is not known. Second, the more biological information we take into account, the more parameters have to be known. However, most of these parameters have never been measured and hence it is very hard, or even impossible to attain their values. This implies that more unknowns are introduced into the more complex models.

In the present models, we incorporate the behavior of the phenotypes fibroblasts and myo-fibroblasts. These cells consume and secrete a chemical signal, which is incorporated in the current models. Furthermore, collagen is incorporated in the models. Since we are interested in the behavior of the dermis in the vicinity of the burn injury, the domain of computation is too large to deal with individual cells and collagen, and therefore we use the cell densities and the density of collagen. As cellular processes, we incorporate migration (by random walk or diffusion and by haptotaxis, which is migration towards the gradient of a chemical through the extracellular material), proliferation, apoptosis and cell differentiation. Fibroblasts differentiate to myo-fibroblasts, which pull with larger forces the extracellular material, and which also produce more collagen than fibroblasts do, under particular circumstances. Since the pulling forces of myo-fibroblasts are much larger than the forces that are exerted by the fibroblasts, we neglect the forces that are exerted by the fibroblasts.

The signaling molecule (chemokine) is supposed to diffuse through the tissue. Furthermore, this molecule is secreted by the fibroblasts and myo-fibroblasts. This molecule, as well as the collagen, is broken down by Matrix Metallo Proteinases (MMPs). This process is taken into consideration in the mathematical model. The production of collagen by fibroblasts and myo-fibroblasts is incorporated into the modeling. Further, we also take into account the orientation of the collagen that is determined according to the migration pattern of the (myo-)fibroblasts.

Next to these issues, a model has been constructed for the mechanical properties of skin. Here, quantities like stiffness and the Poisson ratio play a role. First, a mechanical balance is imposed. Here, one takes into account the pulling forces exerted by the myo-fibroblasts. Further, the orientation of the collagen is taken into account so that the stiffness varies with the collagen orientation. Since a permanent contraction, which can result into a contracture, in which the patient loses mobility, possibly occurs, the model has been adjusted aiming at simulation of permanent deformations. The first simulation results conjecture that a permanent contraction can be modeled, however, rigorous mathematical analysis is needed to formally demonstrate the possibility of simulation of permanent deformations. This issue is incorporated through a morphoelastic description of skin mechanics and this aspect is innovative in our current modeling.

Results

As a first example of simulation results, we show the evolution of a hypertrophic scar over time. Figure 2 shows the distribution of fibroblasts (first row), myo-fibroblasts (second row), signaling chemical (third row) and the collagen density (fourth row) at consecutive times. Next to these profiles, one can see the shape of the skin at these various times. In this simulation, a neo-Hookean description of the mechanical behavior of skin has been used. Shortly after occurrence of the wound, it can be seen that there is an inflow of fibroblasts (see top row). In the early stages myo-fibroblasts are absent, however, in the wound area a

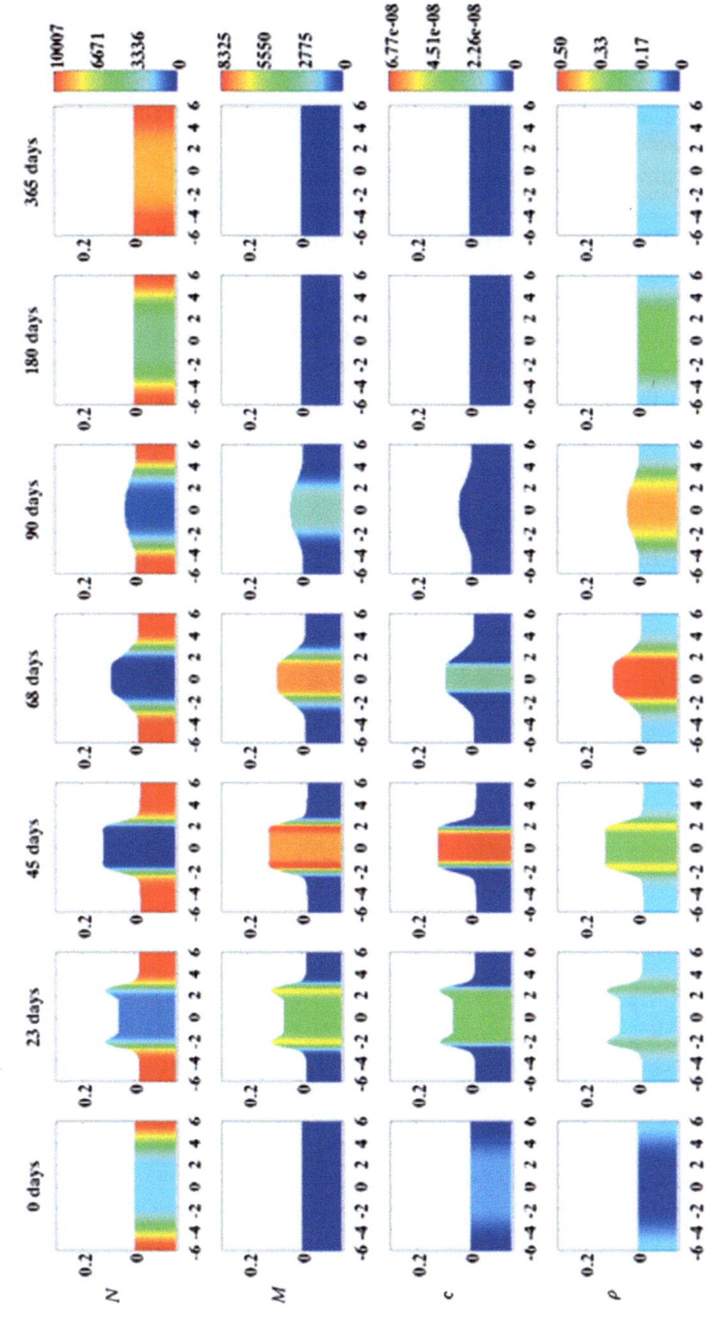

Fig. 2 The distribution of fibroblasts, myo-fibroblasts, chemical signaling molecules, and collagen (from *top* to *bottom* respectively) at consecutive times over a period of half a year (from *left to right* the time increases)

population of fibroblasts develops a bit later as a result of cell differentiation. This has been visualized in row 2. As time proceeds, the number of myo-fibroblasts decreases as a result of apoptosis. Further, it can be seen that the initially present chemical signaling molecules diffuse gradually away from the wound into the rest of the dermal region. Subsequently, these chemical signaling molecules are produced by the (myo-)fibroblasts. Eventually, the concentration of the chemical signal drops down to zero, see the third row in Fig. 2.

Collagen is produced by the (myo-)fibroblasts and at a certain moment the maximum value exceeds the natural equilibrium value. During the final stages of the healing process, the collagen density will decrease as a result of break-down by MMPs. The collagen density will eventually be equal to the natural density that is coupled to undamaged skin, see the bottom row of Fig. 2. Further, it can be seen that during the early stages, the skin shape becomes bulged, which is reminiscent to a hypertrophic scar. The simulations seem to indicate that if the differentiation rate is higher, and if the apoptosis rate is lower, then the myo-fibroblasts will remain longer there, by which the scar, that is the bulged region, will be present over a larger period of time. In the papers of Koppenol et al. (2016a, b), this situation is described in more detail.

Figure 3 shows the area of a skin graft over time. This result for a skin graft has been obtained with a different model, in which the skin has been treated as a morphoelastic material. It can be seen that the wound contracts at the early stages.

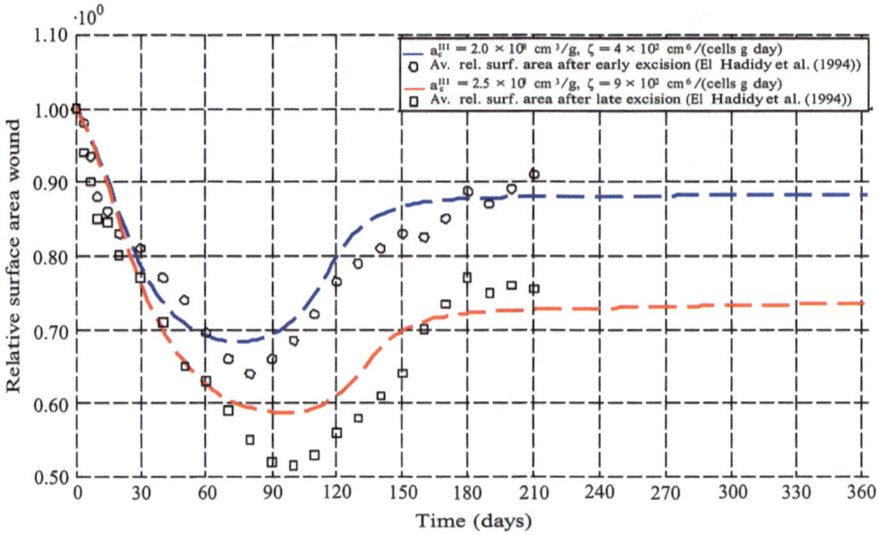

Fig. 3 The area of the skin graft as a function of time. The *circular dots* represent experimental results

This contraction is a result of pulling forces that are exerted by the myo-fibroblasts. In the subsequent stages, the myo-fibroblasts die as a result of apoptosis. Therewith, the pulling forces on the extracellular material disappear. Nevertheless, in Fig. 3, it can be seen that the skin graft no longer deforms back to its original shape. This behavior results since the deformation was so large that this deformation is beyond the elastic region in the stress–strain curve. Now, the skin graft has deformed permanently and this may yield a limitation for the mobility of the patient because of the contracture that may occur. For more information, we refer to Koppenol and Vermolen (2017).

Discussion

The models that we have shown in this manuscript are all based on cell densities and they have a fully continuum-scale character. In order to consider the processes on a very small scale, it could be worth to consider small-scale models like the cellular Potts formulations or semi-continuous cell-based models. However, for larger scales, it is appropriate to use the current partial differential equations-based continuum-scale models since cell-based models would require too many computational resources. Research of interest could be directed to the upscaling of small-scale models to larger scales. The current models further miss the link to the immune system. This link could be investigated more in clinical experiments so that models that couple wound contraction and hypertrophy can be formulated. Of course, a model is a representation of reality, though we try to base the modeling as much as possible on experimental results from several groups. Some caution should be taken since different model formulations may lead to the same results and this may generate some ambiguity in the explanation of experimental results. However, as a tool to forecast the behavior of skin under different circumstances, the models could be helpful in reducing the number of (animal) experiments. It is also believed that cells and skin constitution varies from patient to patient and to this extent, the model results should be subject to a sensitivity analysis. In this sense, a model should not only predict how and whether a contraction or hypertrophic scar develops, but also the likelihood that a scar or contraction develops over time given some initial predefined configuration. This type of information could be of interest to physicians. We believe that regression techniques, stochastic processes, and a statistical evaluation are indispensable in future simulation studies. A first attempt toward this direction has been done in Koppenol et al. (2016). Further, the mentioned models could be combined into one model containing morphoelasticity as well as the ability to predict the occurrence of both hypertrophic scars and contraction by using one single model.

Acknowledgements The authors wish to thank the Dutch Burns Foundation (Nederlandse Brandwondenstichting) for the financial support under project WO/12.103.

References

Koppenol DC, Vermolen FJ (2017) Biomedical implications from a morphoelastic model for the simulation of contracture formation in skin grafts that cover excised burns. Biomech Model Mechanobiol 16(4):1187–1206. doi:10.1007/s10237-017-0881-y

Koppenol DC, Vermolen FJ, Niessen FB, van Zuijlen P, Vuik C (2016a) A mathematical model for the simulation of the formation and subsequent regression of hypertrophic scar tissue after dermal wounding. Biomech Model Mechanobiol. doi:10.1007/s10237-016-0799-9

Koppenol DC, Vermolen FJ, Niessen FB, van Zuijlen P, Vuik C (2016b) A biomechanical mathematical model for the collagen bundle distribution-dependent contraction and subsequent retraction of healing dermal wounds. Biomech Model Mechanobiol. doi:10.1007/s10237-016-082102

Koppenol DC, Vermolen FJ, Koppenol-Gonzalez GV, Niessen FB, van Zuijlen P (2016c) A mathematical model for the simulation of the contraction of burns. J Math Biol. doi:10.1007/s00285-016-1075-4

Vermolen FJ (2016) Particle methods to solve modeling problems in wound healing and tumor growth. Comput Part Mech 2(4):381–399

Investigation of the State of Stress Generated by High Loads in the Ovine Lumbar Intervertebral Disc Using a New Anisotropic Hyperelastic Model

Gloria Casaroli, Fabio Galbusera and Tomaso Villa

Abstract Disc herniation is one of the main causes of low back pain, and it is the pathologic condition for which spinal surgery is most often required. Many experimental and numerical studies have been conducted to investigate the mechanical failure of the intervertebral disc (IVD); however, there is not in the literature a study that defines a mechanical criterion for the disc failure. The aim of this study was to investigate the state of stress generated by the application of high loads and to define which state of stress was the most responsible for herniation. A finite element model of the ovine lumbar IVD was developed. The loading scenarios applied in an experimental study taken from the literature were applied, and the results compared to define the failure conditions. Then the effect of combined and simple rotations was investigated as well. It was found that an axial stress higher than 10 MPa in the posterior region of the annulus has a high probability of damaging the disc, and that flexion had a main role in damaging annulus tissue.

Introduction

Disc herniation is one of the main causes of low back pain, and it is the pathologic condition for which spinal surgery is most often required (Boos and Aebi 2008). In the past years, many experimental and numerical studies have been conducted to investigate the mechanical failure of the intervertebral disc (IVD) (Hansson et al. 1987; Dolan et al. 2013; Schmidt et al. 2007), but to the authors' knowledge, there is not in the literature a study that defines a mechanical criterion for the disc failure. The experimental studies used different animal species and loading conditions (Callaghan and McGill 2001; Fazzalari et al. 2001; Wade et al. 2014); at the same

G. Casaroli (✉) · T. Villa
Department of Chemistry, Material and Chemical Engineering "G. Natta", Politecnico Di Milano, Milan, Italy
e-mail: gloria.casaroli@polimi.it; gloria.casaroli@grupposandonato.it

G. Casaroli · F. Galbusera · T. Villa
IRCCS Istituto Ortopedico Galeazzi, Milan, Italy

A. Gefen and D. Weihs (eds.), *Computer Methods in Biomechanics and Biomedical Engineering*, Lecture Notes in Bioengineering,
DOI 10.1007/978-3-319-59764-5_13

time, the most of the numerical models represented the human disc and have many aspects of complexity due to the composite nature of the annulus fibrosus (AF) (Eberlein et al. 2001; Galbusera et al. 2011; Long et al. 2016). We believe that an anisotropic hyperelastic material is a good alternative for the representation of both the fibers and the ground substance to simplify the disc modeling.

The aim of this study was to investigate the state of stress generated by the application of high loads and to define which mechanical loads and state of stress were the most responsible for disc damaging. A finite element model of the ovine lumbar disc was generated and anisotropic hyperelastic properties were assigned to the AF (Casaroli et al. 2016). Then the model was used to investigate the effect of the application of high loads conditions (Casaroli et al. 2017). Finally, the results were combined with the outcomes of a parallel in vitro study (Berger-Roscher et al. 2016) to define a mechanical criterion for the annulus failure.

Materials and Methods

A finite element model of the ovine lumbar IVD was developed. The geometry of the disc was generated by the reconstruction of the caudal and cranial vertebral bodies of a L3-4 segment (Avizo 8.0) using a specific Python script. The AF was described as an anisotropic hyperelastic material, whereas the mechanical properties of the nucleus pulposus (NP) and of the cartilagineous and bony endplates (CEP and BEP, respectively) were taken from the literature (Table 1). The anisotropic behavior was described by the Holzapfel–Gasser–Ogden formulation and implemented in Abaqus 6.14-5, and the parameters were defined using a full factorial optimization method. C_{10} and D described the ground substance and were defined combining in vitro uniaxial compressive tests (Little et al. 2010) with numerical

Table 1 Material properties assigned to the disc structures

Structure	Material behavior	C_{10} (MPa), D (MPa^{-1})	K_1 (MPa), K_2 (MPa), κ	References
Anterior AF	Anisotropic hyperelastic	0.0605, 0.311	24, 1700, 0.01	Casaroli et al. (2016)
Lateral AF	Anisotropic hyperelastic	0.0327, 0.615	5, 940, 0.01	Casaroli et al. (2016)
Posterior AF	Anisotropic hyperelastic	0.0772, 0.261	1, 50, 0.01	Casaroli et al. (2016)
NP	Neo-Hookean	0.1678, 0.120	–	Ayturk et al. (2012)
		E (MPa), υ		
CEP	Linear elastic	24, 0.4	–	Ayturk et al. (2012)
BEP	Linear elastic	1000, 0.3	–	Ayturk et al. (2012)

simulations. K_1, K_2 and k described the influence of the collagen fibers and were established based on tensile tests on annulus specimens and numerical simulations. For the sake of validation, the flexibility of the model was measured in flexion-extension (FL-EX), lateral bending (LB), and axial rotation (AR), and the range of motion compared with the literature (Reitmaier et al. 2014).

The model was then constrained at the caudal bony endplate in all degrees of freedom, and five different complex loading conditions were investigated. In the worst case an axial compression (AC) of 800 N was applied together with 4° of AR, 10° of LB and 13° of flexion FL. In the other four cases only three of these conditions were applied. The applied rotations were double of the physiological ones described by Reitmaier et al. (2014). The simulations were performed in Abaqus explicit. It was assessed that the energy balance was respected. Three sections in the posterolateral regions where failures experimentally occurred were defined, and the averaged principal stresses and strains in the circumferential, radial and axial direction were measured. A multiple linear regression analysis based on the combination of the experimental and numerical outcomes was performed to investigate which stresses were predictive for the disc failure. Thresholds defining a high and a low risk of failure were defined.

Then, the effect of each pair of rotations and of the combination of a single rotation with axial compression was investigated. The stresses were analyzed and compared to the defined thresholds. Finally, the effect of the single rotation and axial compression were analyzed as well.

Results

The values of the parameters that described the anisotropic hyperelastic behavior of the AF (Table 1) showed a good fit between the numerical and the experimental results. The flexibility of the disc was in the range defined by the literature (Reitmaier et al. 2014). The numerical simulations showed that when AC and all the rotations were applied the stress was the highest. In the circumferential direction, the stress state was highest in the anterior region of the AF, whereas in the axial and in the radial direction it was highest in the posterocontrolateral region (Fig. 1). Stress distribution was not uniform from the caudal to the cranial and from the inner to the outer part of the annulus. In the posterolateral region, the stress was highest close to the endplates in the circumferential and in the radial direction, and in the inner region of the annulus in the axial direction (Fig. 2). When FL and LB were applied, the state of stress was the highest in the axial and in the circumferential directions, whereas AC did not increase the state of stress in the tissue. The statistical analysis revealed that the axial and circumferential stress located in the caudal and in the middle part of the annulus was predictive of disc failure. When AC and all rotations were applied, the axial and the circumferential stress was up to 12 and 10 MPa respectively, whereas when FL was not applied the stress was lower than 4 and 6 MPa. Axial stress thresholds defining a high and a low risk

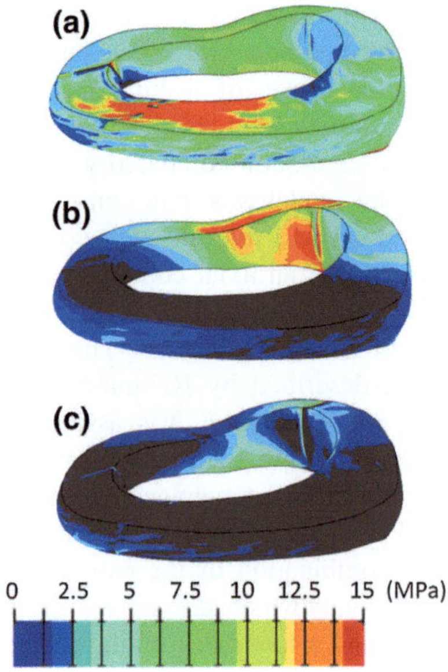

Fig. 1 Stress distribution in the AF when AC and all rotation were applied in the circumferential (**a**), axial (**b**) and radial (**c**) direction. The *gray* regions represent the parts in which the stress was negative

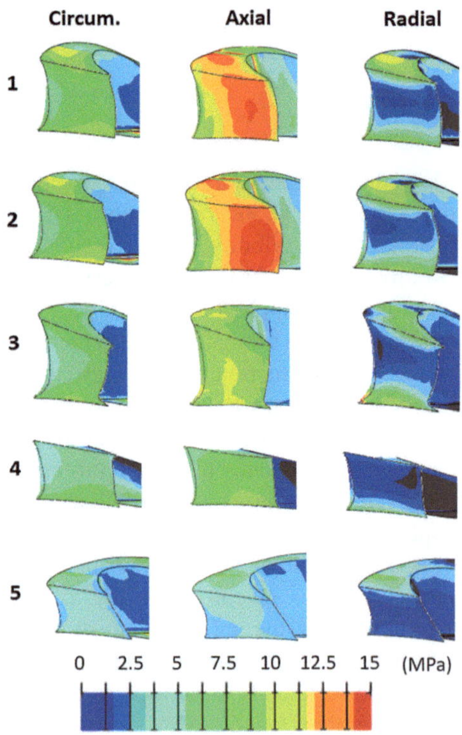

Fig. 2 Stress state distribution within the posterolateral section of the AF in the circumferential (circum.), axial and radial direction in the studied loading scenarios (1–5). The *gray regions* represent the part in which the stress was negative

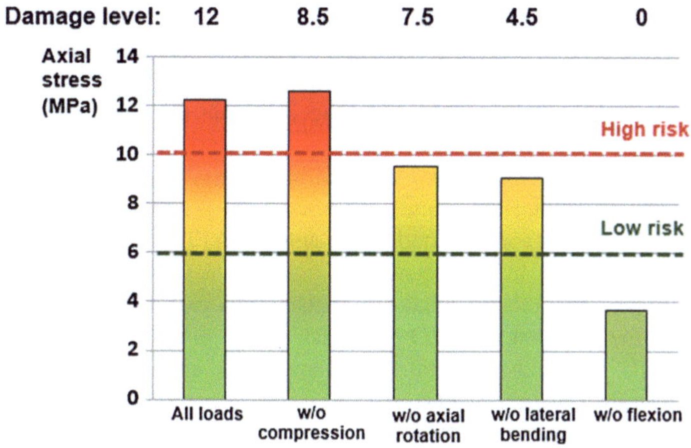

Fig. 3 Axial stress generated in the PL-1 section in the caudal region in the five loading conditions. In the *upper part* the recurring of AFF is indicated, whereas in the *lower part* the risk failure of the group is written

of failure were defined and fixed to 10 and to 6 MPa, respectively (Fig. 3). In the circumferential direction they were fixed to 9 and 6 MPa.

The application of pair of loads, showed that when FL was combined with another rotation the state of stress was the highest in all the directions, and a high risk of failure was generated only by the application of FL and AR. Finally, when single rotation was applied, only FL generated a moderate risk of failure (7 MPa).

Discussion

In this study, a finite element model of the ovine lumbar IVD with anisotropic hyperelastic material has been generated. The use of an anisotropic formulation had two advantages: it allowed simulating the presence of the collagen fibers without modeling them, keeping the mesh refinement easy, and it was based on only five parameters that were experimentally definable. Different material properties were assigned to the different regions of the AF. The flexibility of the model was in the range described in the literature, although it overestimated the average values. A possible explanation of this effect could be the geometrical dimensions.

Then, the model was used to investigate the mechanical tests conducted in a parallel in vitro study and a predictor of the risk of failure of the IVD has been proposed. The numerical results were in agreement with the experimental ones, demonstrating that complex loading conditions are responsible for disc failure. When FL was not applied the stress in posterior region of the annulus was lower than in the other cases, as well as the in vitro study has shown. The analysis of the stress state generated by less complex scenarios showed that the combinations in which flexion was included generated a high risk of failure. Moreover, the analysis of the application of single rotations supported the conclusion that flexion had a main role in damaging annulus tissue. Indeed, according to the defined limits for the disc damage, FL alone can cause disc damage, whereas the other rotations keep the annulus in a safe condition.

The highest state of stress was generated in the axial and in the circumferential directions, which defined the plane of the collagen fibers. In the circumferential direction the stress was concentrated in the anterior AF, but it was supposed to not be responsible for disc failure because it was caused by the compression of the elements: in fact, the related axial stress was negative, reducing the real stress supported by the collagen fibers. On contrast, in the posterolateral region the axial stress was high in both the circumferential and in the axial direction, ensuring the tensile state of the collagen fibers.

In conclusion, this study presents a model and its application that allows investigating the effect of any loading conditions on the disc and estimating the risk of generating failures.

References

Ayturk UM, Gadomski B, Schuldt D, Patel V, Puttlitz CM (2012) Modeling degenerative disk disease in the lumbar spine: a combined experimental, constitutive, and computational approach. J Biomech Eng (134):101003

Berger-Roscher N, Casaroli G, Rasche V, Villa T, Galbusera F, Wilke H-J (2016) Influence of complex loading conditions on intervertebral disc failure. Spine 42(2):E78–E85

Boos N, Aebi M (2008) Spinal disorders. Fundamentals of diagnosis and treatments. Springer, Berlin

Callaghan JP, McGill SM (2001) Intervertebral disc herniation: studies on a porcine model exposed to highly repetitive flexion/extension motion with compressive force. Clin Biomech 16:28–37

Casaroli G, Galbusera F, Jonas R, Schlager B, Wilke H.-J, Villa T (2016) A novel finite element model of the ovine lumbar intervertebral disc with anisotropic hyperelastic material properties. PLoS ONE 12(5):e0177088

Casaroli G, Galbusera F, Jonas R, Schlager B, Wilke H-J, Villa T (2017) Numerical prediction of the mechanical failure of the intervertebral disc under complex loading conditions. Materials 10(1):31

Dolan P, Luo J, Pollintine P, Landham PR, Stefanakis M, Adams MA (2013) Intervertebral disc decompression following endplate damage. Spine 38(17):1473–1481

Eberlein R, Holzapfel GA, Schulze-Bauer CAJ (2001) An anisotropic model for annulus tissue and enhanced finite element analyses of intact lumbar disc bodies. Comput Methods Biomech Biomed Engin 4(3):209–229

Fazzalari NL, Costi JJ, Hearn TC, Fraser RD, Vernon-Roberts B, Hutchinson J, Manthey BA, Parkinson IH, Sinclair C (2001) Mechanical and pathologic consequences of induced concentric anular tears in an ovine model. Spine 26(23):2575–2581

Galbusera F, Schmidt H, Neidlinger-Wilke C, Gottschalk A, Wilke H-J (2011) The mechanical response of the lumbar spine to different combinations of disc degenerative changes investigated using randomized poroelastic finite element models. Eur Spine J 20:563–571

Hansson TH, Keller TS, Spengler DM (1987) Mechanical behavior of the human lumbar spine. II. Fatigue strength during dynamic compressive loading. J Orthop Res 5:479–487

Little JP, Pearcy MJ, Tevelen G, Evans JH, Pettet G, Adam CJ (2010) The mechanical response of the ovine lumbar anulus fibrosus to uniaxial, biaxial and shear loads. J Mech Behav Biomed Mater 3:146–157

Long RG, Torre OM, Hom WW, Assael DJ, Iatridis JC (2016) Design requirements for annulus fibrosus repair: review of forces, displacements and material properties of the intervertebral disc and a summary of candidate hydrogels for repair. J Biomech Eng 138:26720265

Reitmaier S, Volkheimer D, Berger-Roscher N, Wilke H-J, Ignatius A (2014) Increase or decrease in stability after nucleotomy? Conflicting in vitro and in vivo results in the sheep model. J R Soc Interface 11:20140650

Schmidt H, Kettler A, Rohlmann A, Claes L, Wilke H-J (2007) The risk of disc prolapses with complex loading in different degrees of disc degeneration—a finite element analysis. Clin Biomech 22(9):98–988

Wade KR, Robertson PA, Thambyah A, Broom ND (2014) How healthy disc herniate: a biomechanical and microstructural study investigating the combined effects of compression rate and flexion. Spine 39(13):1018–1028

Soft Tissues' Loadings on Healthy Knee at Different Physiological Flexions: A Coupled Experimental–Numerical Approach

Boris Dousteyssier, Jérôme Molimard, Chafiaa Hamitouche, Woo-Suck Han and Eric Stindel

Abstract In this study, the movement of climbing a step is decomposed in 4 EOS images. A patient-dependent 3D model of the knee is then created from MRI, and several numerical simulations are carried out according to the experimental boundary conditions (force and flexion angle), so as to ensure the global knee mechanical equilibrium. To validate this patient-specific model, its bony structure is confronted with the EOS images once the mechanical equilibrium is reached. This model gave us an estimation of the stress in the ligaments for every flexion angle as well as a pressure map on the cartilages.

Introduction

Knee osteoarthritis is one of the major causes of disability in older adults. According to a study conducted in the United Kingdom and the Netherlands on a population of 55 y.o. and more, 25% declared having chronical and persistent knee pain (Peat et al. 2001). Osteoarthritis is the combined effect of degradation of the cartilage, thickening of the subchondral bone and new bone formation on the edges of the cartilages. The main clinical diagnosis tool for osteoarthritis is the radiography, with effects being apparent on X-ray images: joint space narrowing for the cartilage destruction, variation of the signal intensity for the thickening of the

B. Dousteyssier · J. Molimard · W.-S. Han (✉)
CIS-EMSE, Ecole Nationale Supérieure Des Mines de Saint-Etienne,
42023 Saint Etienne, France
e-mail: han@emse.fr

B. Dousteyssier · J. Molimard · W.-S. Han
INSERM U1059, SAINBIOSE, 42023 Saint Etienne, France

B. Dousteyssier · J. Molimard · W.-S. Han
Université de Lyon, 69000 Lyon, France

B. Dousteyssier (✉) · C. Hamitouche (✉) · E. Stindel
Laboratoire de Traitement de l'Information Médicale, INSERM UMR 1101,
29609 Brest, France

A. Gefen and D. Weihs (eds.), *Computer Methods in Biomechanics and Biomedical Engineering*, Lecture Notes in Bioengineering,
DOI 10.1007/978-3-319-59764-5_14

subchondral bone and the bone formation is directly seen. The knee degradation and the associated pain when developing osteoarthritis are strongly related not only to the pressure on the cartilage, but also to the knee stability and to the subsequent loadings on the ligaments. Numerical models of the knee can help us evaluate both of these in a daily situation for the knee and understand the osteoarthritis phenomenon. Some knee joint models already exist in the literature. Some authors proposed knee models without any flexion (Pena et al. 2006; Donahue et al. 2002). But those models as incomplete as they may be only represent a small part of the knee functionality, and most likely not the part that is significant in the deterioration of the knee. Other authors have introduced knee flexion, for example by directly controlling active muscular groups (Marouane et al. 2015). But the pressure applied on the articulations may be questioned, given the difficulty to assess the physiological level of the forces used to perform the flexion. On another paper, the knee flexion is modelled by using a full displacement-controlled model (Zhong et al. 2011). In this case the uncertainties of the bone positioning may lead to high variations in the cartilage pressure and ligament loads. It especially removes the link between the two phenomena, separating the cartilages and the ligaments by the bones that are completely controlled by the user.

Here our approach is coupled, using both the experimental data by the means of medical imaging and a finite element model to have a physiological representation of the knee and its solicitations at a given flexion angle. In order to do so, a geometrical model is created from the MRI, with which a numerical simulation of an experimental setup is conducted. The experimental setup modelled is the EOS imaging of the movement decomposition of climbing a step. The EOS is a low-dose X-ray system (Wybier and Bossard 2013) and thus is already used in order to diagnose knee osteoarthritis. Its acquisition area allows upright position during the acquisition, as well as some room to instal a step. An added benefit of the climbing a step movement is that the flexion angle of the knee is higher than during the gait movement (Kaufman et al. 2001). The goal of this approach, a numerical simulation of an experimental setup, is to create a model that will be validated by comparing its numerical results with the EOS data. This model will reveal the roles of the ligaments during the knee flexion and give pressure maps on the cartilages.

Method

EOS Acquisition

A healthy volunteer (24 y.o., BMI = 26.23 kg/m^2) undergoes an EOS imaging of the right knee decomposing the movement of climbing a step. The height of a step ranges from 130 to 180 mm or even 210 mm for stiff stairs. For practical reasons, this study was conducted for a step of 150 mm. Between each acquisition, a 50 mm thick block is added under the foot of the volunteer until the step height of 150 mm is reached. During the whole procedure, a weighting scale is positioned under the

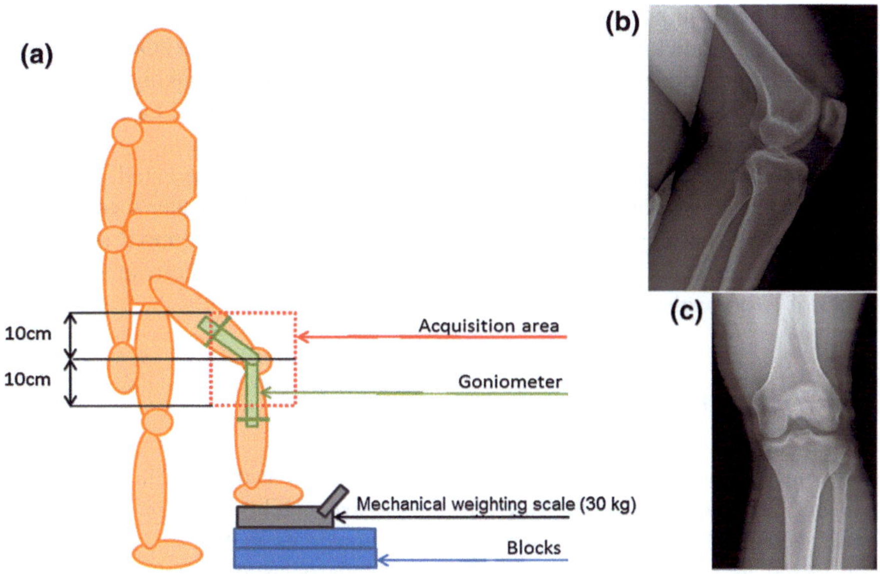

Fig. 1 **a** Schematics of the EOS acquisition setup—side view. **b** Sagittal scan of the knee at 55° of flexion—i.e., 10 cm height step. **c** Matching frontal scan of the knee

foot of the volunteer for him to maintain a constant load of 30 kg. This ensures us the most homogenous boundary conditions possible on the knee at different flexion angles for the following numerical simulations. This load has been chosen to be the most comfortable to maintain while being immobile during the whole acquisition procedure. Even though the EOS acquisition is rather quick, approximately 10 seconds per flexion angle, a lighter load has proven to be difficult to control at lower flexion angles and respectively a heavier one is near impossible for higher flexion angles. The data obtained is 5 sagittal and 5 frontal scans of the knee hard tissues at the flexion angles of 0°, 40°, 55°, and 70°. Those images give access to the physiological position of the bones at those specific angles of flexion. The schematics of this acquisition and examples of data obtained are shown in Fig. 1.

3D Geometric Model

The 3D geometrical model of the bones and cartilages used in the simulations to be as complete and physiologic as possible, it is segmented from three different MRI stacks of the volunteer's knee. The MRI modalities used are SPIN echo for the cruciate ligaments insertions, SPIN echo with fat saturation for the lateral ligaments insertions and the cartilages and finally gradient echo for the bones of the articulation. These three MRI stacks has to be fitted one on the other in order for all the different segmented parts to be coherent in the model. The software used for the segmentation is AVISO® (FEI, Hillsboro, United States). The segmented model is then smoothed with a low-pass filter using the toolbox GIBBON® for MATLAB® (MathWorks, Natick, United States) and meshed with HARPOON® (Sharc, Manchester, United Kingdom). Those different steps in the making of the geometrical model can all be source of uncertainties and so a special care is taken so that the patient-specific aspect of this model is not lost during the process. The study on the tools used for the model creation has been presented previously (CMBBE-2015) and only a brief recall is given here for the reader's comprehension. It shows that the error between the final geometry and the initial data is less than a millimetre, which is a very good precision because it corresponds to the width of two voxels in the MRI raw data.

Finite Element Modelling

In the present model, only the tibia and the femur are considered. The patella is not included because it can be neglected for the passive stability-oriented analysis. The bones being a lot more rigid than any other materials in the analysis, they are considered as rigid bodies. The cartilages are viscoelastic tissues; however this analysis being static, the viscosity has no impact in the equilibrium of the model. And thus they are modelled as single-phase linear elastic and isotropic material with an elastic modulus of $E = 12$ MPa, and a Poisson ratio of $\nu = 0.45$ (Yao et al. 2006). The cartilages are tied to their respective bones, and the contact between the two cartilages is defined as "hard", meaning that no intersection whatsoever is allowed between the two surfaces in contact. A friction coefficient of $\mu = 0.08$ has been taken. Finally the ligaments modelled are the anterior cruciate ligament (ACL), the posterior cruciate ligament (PCL), the lateral collateral ligament (LCL), and the medial collateral ligament (MCL). They are modelled as truss elements working in tension only, and pined on the segmented insertions area. As truss elements they are considered as transversely isotropic. The cross-sectional areas for the ligaments are 42, 60, 18, 25 mm^2 for respectively the ACL, PCL, LCL and MCL. Longitudinally, they are modelled as hyperelastic materials following the stress/strain curve given by Gardiner and Weiss (2003). In the standing position all ligaments have an initial tension of 100 N (Oh et al. 2014), this initial tension has been adjusted to the differences in strain of the different flexion angles and then transferred accordingly. The inferior insertion point of the LCL is on the fibula. But since the bones are modelled as rigid bodies, the fibula does not have any other relevant role in this simulation. So to lighten our model, we simply add an element on the tibia in order to reach the correct LCL insertion on the fibula. The geometrical model is shown in Fig. 2. It is composed of 43,500 linear elements (tetrahedron, hexahedron and prism). During the convergence analysis, it is noticed that the number of elements in the bones or ligaments matters little due to their modelling. The bones modelled as rigid bodies only require enough element for their contact surface to be smooth. Otherwise, it would create stress concentration on the cartilages on the tied nodes. The ligaments taken individually are in a very

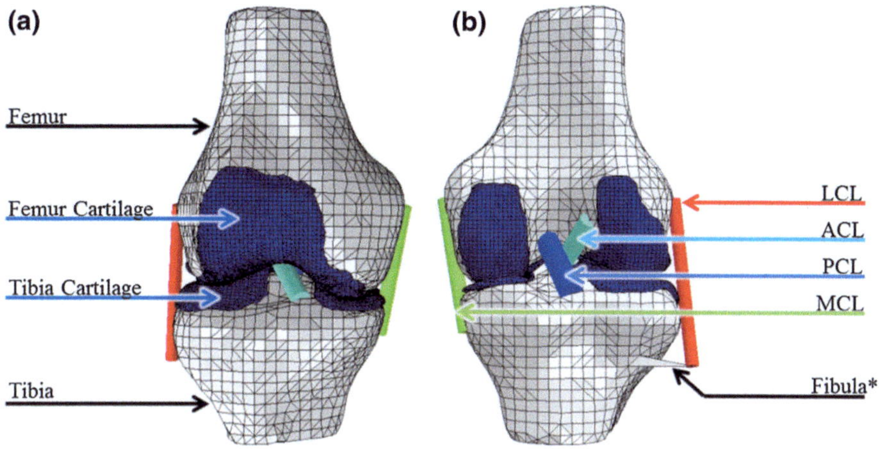

Fig. 2 Finite element model of the knee joint. **a** Front view. **b** Back view. *Asterisk* element added on the Tibia to reach the physiological inferior insertion point of the LCL

simple state. Since no contact are defined other than their insertion areas, they work as a hyperelastic spring. For those reasons most of the elements are in the cartilages, the most critical parts of this model: they make for 29,500 of the elements.

Numerical Simulations

The simulations are carried on ABAQUS® (Dassault Systèmes, VélizyVillacoublay, France). The goal of each simulation is to respect the experimental conditions of the EOS data acquisition. In order to do so, the initial bone position of this model is fitted on the physiological positions from the EOS data for the four different flexion angles. We managed to fit the 3D model on two 2D EOS images so as to check the contour of a projection of the 3D model, then to fit and to compare it with the segmented contour of the EOS images. The method used is based on the Iterative Closest Point (Zhang et al. 2016) in order to fit the two contours, and use the simplex method to find the best match between the possible bone contours from the model with the segmented contours of the EOS images. The fitting precision obtained is 1 mm on the two projected planes and 0.5° along the bone axis. The specifics of the algorithm and an example of contour fitting are shown in Fig. 3.

Once the bones are in their physiological positions for each flexion angles, an intersection between the cartilage of the femur and those of the tibia can be noticed. The numerical simulation is composed of the following steps: (1) slightly move the tibia downward to remove this intersection, while keeping the femur blocked, (2) put the boundary conditions of the experimental setup, (3) obtain the global knee mechanical equilibrium by applying the same load as the experimental

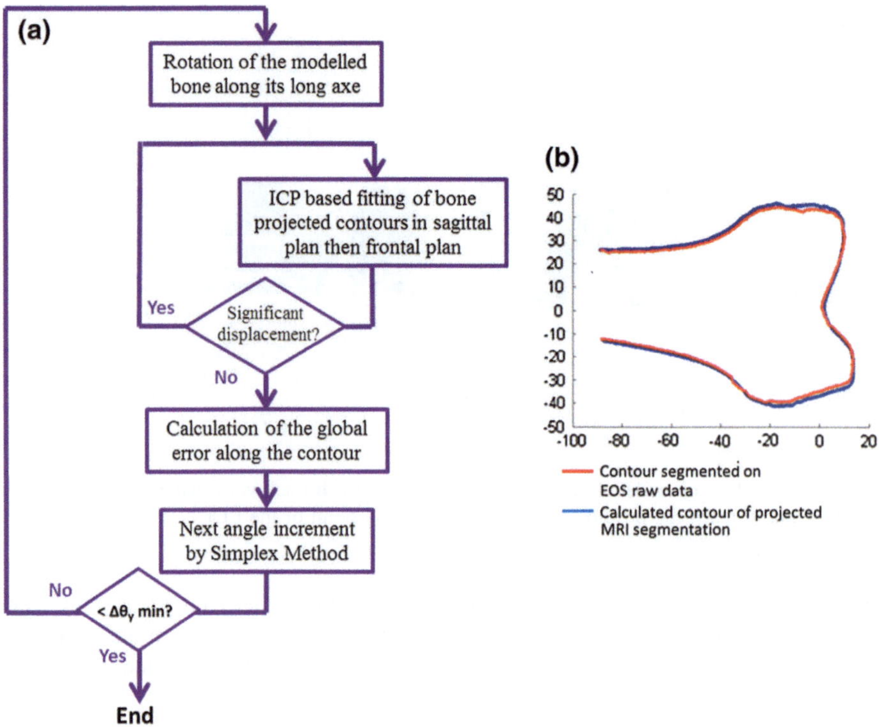

Fig. 3 **a** Fitting algorithm of a 3D model on two 2D projections. **b** Example of a contour comparison: frontal plan of the femur at a 0° flexion angle

setup. All along the analysis, the femur is fixed in space; all degrees of freedom of the tibia are initially locked too. The first step consists in translating the tibia along its axis until there is no more intersection between the cartilages. At this point the contact between the cartilages is implemented and the force on the tibia is added: 300 N along its axis. Then all degrees of freedom are released except the rotation in the sagittal plane that is the flexion angle. The FEA is carried out until the global knee mechanical equilibrium. Once achieved, the equilibrium position of the tibia is compared with its initial position, directly linked to the experimental data, to assess the validity of the model. All steps of the FEA are summarized in Fig. 4.

Results

Tibia Position

The first result sought is the tibia position. Comparing this position with the initial EOS data ensured that the model and the global equilibrium reached are indeed

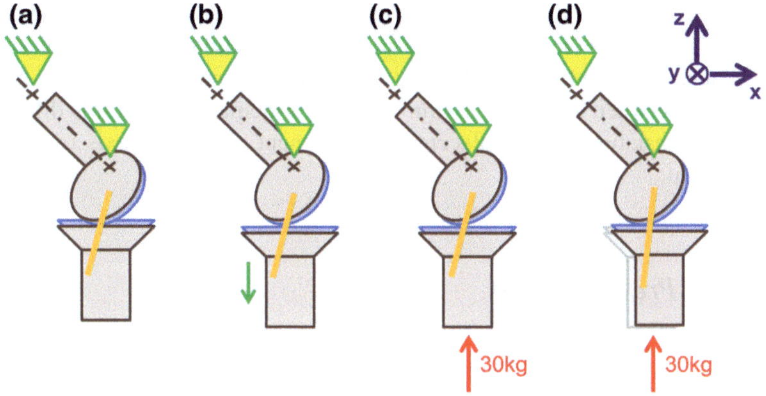

Fig. 4 FEA steps: **a** initial position, **b** removal of the cartilages intersection, **c** addition of the contact and experimental boundary conditions, **d** global knee mechanical equilibrium

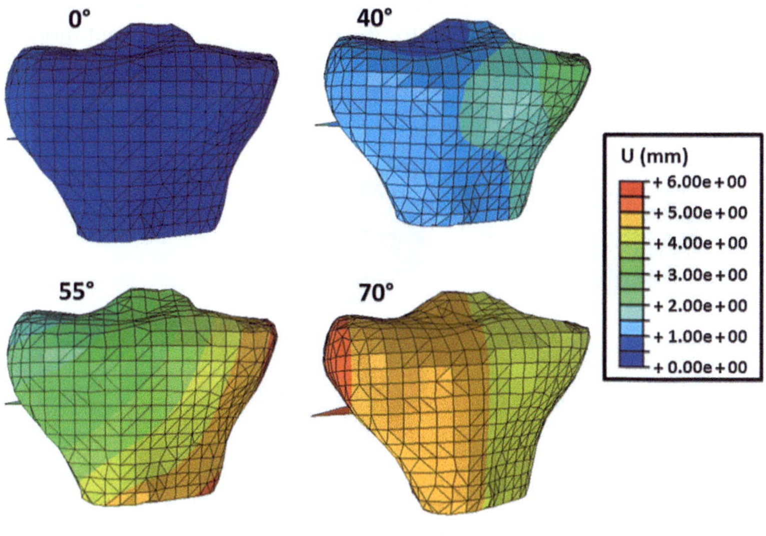

Fig. 5 Position distance of each node to their counterpart on the experimental data

physiological. On Fig. 5, we can see the distance of each node from their experimental position for the four flexion angles studied, 0°, 40°, 55° and 70°. The simulation at a flexion angle of 0° shows an insignificant distance between the simulation and the experiment for every points of the tibia. The maximum distance observed stands only at 0.7 mm. For the flexion angle of 40°, we can also observe that this distance is very small: for 97% of the nodes, their FEA position is at most 2 mm apart from their EOS position. The greatest distance observed is 2.66 mm. With the gradient of displacement, we can observe that the difference in positions is due to a rotation along the bone axe, measured at 2.6°. On the 55° flexion angle simulation, the distances vary from 1.75 to 5.24 mm, but less than 4 mm for 81% of the nodes. Once again those differences are due to a rotation, this time mainly on the frontal plane, measured at 3.7°. Finally for the flexion angle of 70°, the distances observed are between 4.0 and 5.2 mm. Here the principal cause of this difference is an anterior translation, the displacement component along this axis ranging from 3.8 to 5.0 mm.

Cartilages Pressure Map

In Fig. 6 are shown the different pressure maps for the four simulations. We can see that in each case the pressure is rather balanced on the two condyles, with a slight overpressure on the external condyle. Furthermore the point of maximum pressure follows the line of contact described in other studies (Shandiz et al. 2016). The maximum pressure for the different flexion angles are 3.43, 5.34, 5.89 and 5.23 MPa for respectively 0°, 40°, 55° and 70°. We can also notice that the area under pressure is a little more spread for the flexion angle of 0°, and more focused for 40°, 55° and 70°.

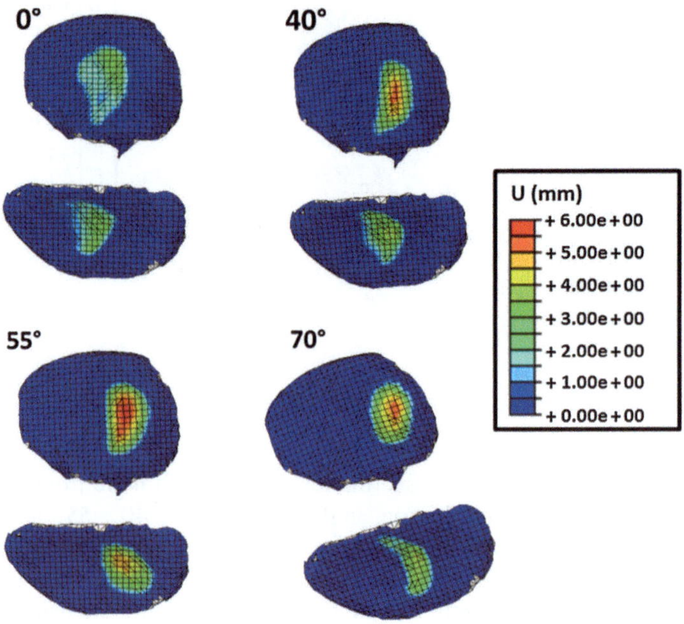

Fig. 6 Pressure map on tibia cartilages after global knee equilibrium

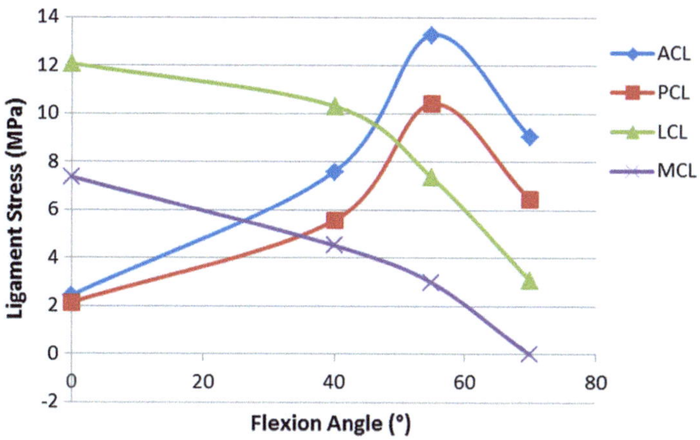

Fig. 7 Stress in the ligaments depending on the flexion angle

Ligament Stress

Finally the last result that is focused on is the ligaments stress repartition depending on the flexion angle. As shown in Fig. 7, we can see that at the lowest flexion angle the collateral ligaments withstand more stress, with their maximum stress at a flexion angle of 0° being 12.07 and 7.37 MPa for the lateral and medial collateral ligaments respectively. Then the stress steadily decreases as the flexion angle increases until the LCL reaches 3.07 MPa and the MCL is totally relaxed for the 70° flexion angle. At 40° flexion angle, the stress is balanced between the different ligaments, but for higher flexion angles it is focused in the cruciate ligaments. The stress in the cruciate ligaments starts lightly at 2.43 and 2.13 MPa for the anterior and posterior cruciate ligaments respectively in the standing position to reach a peak at the 55° flexion angle with stresses of 13.27 and 10.41 MPa.

Discussion

Boundary Conditions and Material Properties

This first patient dependant knee joint model simulate the climbing a step movement. It decomposes it in several quasi static situations at flexion angles ranging from 0° to 70°. It is built combining the EOS imaging system and an MRI acquisition of the volunteer. The MRI data is the base of the geometrical model and the EOS data brought the physiological bone positions at those different flexion angles. This model focuses on the passive stability of the knee and so the modelled parts of the knee are the bones (femur and tibia as rigid bodies), the cartilages (as isotropic elastic and linear elements) and the ligaments (ACL, PCL, MCL and LCL as hyperelastic truss elements). In this first simulation the influence of the linear elements of the cartilages is not yet assessed. A future step will be to use quadratic elements and see if the results differ enough to justify the added simulation time. The contact between the cartilages is defined as hard (with no intersections) with a friction coefficient μ = 0.08. This friction coefficient is rather high but it is the closest possible to the literature [μ = 0.02 (Mow et al. 1997)] in order for all the simulation to reach equilibrium. The limiting simulation is for a flexion angle of 55°. In this simulation the static equilibrium cannot be reached as the tibia seams to slide on the femur. This is a clear limitation of the model that will be addressed in the future. During the numerical simulation, the cartilage intersection due to the bone positioning is removed, the experimental boundary conditions are added and the global knee equilibrium is reached. On average, each simulation took approximately 1100 s of CPU time, which took around 30 min of wall clock time. Even though the positions of the bones are accessible with the EOS images, those simulations cannot be executed simply by controlling the displacement of the bones. When they are done this way most of the high flexion angles simulations do not reach equilibrium: the elements in the cartilages are too much distorted and the simulation end prematurely. At low flexion angle the simulations do reach equilibrium but the results, especially the pressure maps, are not reliable. The resulting force on the tibia is calculated as high as 3200 N, which is more than 10 times the expected result. We can also see that there is a strong imbalance between the pressures on the two condyles: for example at 40° flexion angle, the maximum pressure is 8.7 MPa on the exterior condyle and 5.9 MPa on the interior one.

FE Model Validation by Bone Position

After the creation of the FE model and once the simulations are complete, the first step is to come back to the experimental data, i.e., EOS images, in order to make sure that this model is physiologically correct. The physiological position of the bones in the knee joint given by EOS imaging is compared with their position obtained by FEA after the global knee equilibrium is reached with the same boundary conditions. The numerical results have a good agreement with the EOS images. For the two lowest flexion angles they are as good as they can get: the precision of the fitting of the 3D model on the EOS raw data is precise within 1 mm, meaning that after the equilibrium, the difference on the tibia (only moving part of the model) of 2 mm is within that precision range of EOS data's precision. Thus in that distance interval the difference between the simulation and the experimental data can equally be explained by (1) the simulation equilibrium missing the physiological position by a tiny amount or (2) our position measurement of the experimental data that could slightly be improved. As a reminder, this precision corresponds to only two voxels in the data gathered on the MRI. Therefore, the simulation of the standing leg with a maximal distance of 0.7 mm is totally in its physiological position and the simulation at a 40° flexion angle, with 97% of its nodes at less than 2 mm of their experimental counterparts, is almost perfectly aligned with the experimental data. For the 55° flexion angle simulation, the final position of the bones is still very good; 81% are at most at twice the precision range, making it an acceptable error of position. However, this model can be questioned for the simulation at the flexion angle of 70°. In this state, every node is at least at 4 mm of their aimed position. But in this particular case, we underlined that this distance is mostly due to an anterior translation of the tibia. At this flexion angle, the tibia seams to start to slip more than its physiological movement given by EOS images. The height of a step varying from 130 to 210 mm, we conducted the entire protocol for a fifth flexion angle, for the case of a high step. An EOS image with a 200 mm high step has been taken and the corresponding numerical simulation is launched. But it could not achieve equilibrium, as the minimal time increment is not enough however small it is. On the incomplete result we can observe an anterior translation, similar to the simulation at the 70° flexion angle. It is very likely that at this high flexion angle the tibia starts to slip on the femur and that the ligaments are not enough to stabilise the modelled knee joint.

Model Shortcomings

This model may prove to be lacking some stability tools, like the menisci that would directly be in the way of such a translation. Considering that, adding the menisci is the next step for the development of this model. In two of the simulations, we could also notice a slight rotation of the bone along its long axis. This may be due to the lack of contact between the ligaments and their surroundings. As the model is, nothing stops the cruciate ligaments to intersect each other, whereas physiologically they would twist one on the other and impede this rotation. However, this kind of rotation is very small and to attend to it may not be essential considering how this contact may make the model more complex. Adding contact conditions between the ligaments and the bone can also slightly change the way the lateral collateral ligament would work; adding a small lever arm on its tibia intersection and maybe making the whole knee joint a bit more stable. Once again the changes would prove to be to smalls to make this addition in the model a priority. Finally, the higher the flexion angle, the more the active stability may be impactful. This model focuses in the passive stability but maybe some active stability is necessary for a global equilibrium at higher flexion angle, especially in this case are the knee is loaded. As the patella can be segmented on EOS images, a future model could involve a fitted patella, fixed in space as the femur, linked to the tibia with the patellar tendon.

Pressure Maps and Ligament Stress

This first model, even incomplete, still shows some physiological behaviour. Beside the good bone position, we can see on the pressure maps that the points of highest pressure are in agreement with the contact path described by Shandiz et al. (2016). The pressure is rather well balanced between the two condyles, as expected from a healthy knee. We can separate two kind of pressure map along the flexion; first the standing leg and second every bent knee. In the first case, at the 0° flexion angle we can see that the area under pressure is more spread than the other case and that the maximum pressure is a lot lower, standing at 3.43 MPa. Whereas for the flexion angles of 40°, 55° and 70° it is more focused with a maximum pressure around 5.5 MPa. This dichotomy in the pressure maps is explained by the contact surface of the femur condyles. At a 0° flexion angle the curvature angle of the contact surface is more obtuse, allowing a better stability for the knee in the standing position, making holding it for a long period of time more comfortable. For the other flexion angles, this curvature angle is more acute, making the flexion movement easier, and this movement smoother in general. Moreover, the slight overpressure on the external condyle showcases the different purpose of the two condyles: the internal condyle guiding the flexion while the external sustain a higher load. The stress in the ligaments also highlights the different functions of the ligaments. The collaterals ligaments guarantee the knee joint stability in extension while the cruciate take over in flexion. We can notice that usually it is the medial collateral ligament that is more loaded than the lateral (Dufour and Pillu 2007); in this case it is the opposite because the volunteer suffer from a light varus (4.5°). A maximum stress in the cruciate ligaments along rotation is recorded near 55°. It is due to the complex kinematics of the femur on the tibial plateaux, mixing translation and rotation. Indeed, this value remains qualitative, considering the angular discretization step.

Conclusion

In order to study on the question of knee degradation and pain when developing osteoarthritis, related to both the pressure on the cartilages and the knee stability, a knee joint model was created. We proposed a mixed approach, both using medical imaging (MRI, EOS X-ray system) and force platform in conjunction with a FE model. The goal was to obtain pressure maps on the cartilages of the knee and the stress sustained by its ligaments during a daily activity.

The movement of climbing a step was chosen for its higher flexion angle of the knee joint. A FE model was created, focusing on passive stability and recreating an experimental setup: the decomposition in 4 static EOS images of the movement of climbing a step. In order to do so, a geometrical model of the volunteer's knee was fitted on the physiological bone position obtained on the EOS images. Once the initial parasitic intersection removed, the experimental boundary conditions were added and the numerical simulation is carried on until a global knee mechanical equilibrium was reached. Then the simulated position of the bones could be compared with the experimental one, and we obtained pressure maps on the cartilages and the stress in the ligaments for this controlled situation.

Thanks to this method allowing to compare the FE model with the experimental data, here the bone positions, we know when we can be confident in the results obtained and when we have to be cautious. Here this very simplified model showed that for low flexion angles it could achieve a global knee mechanical equilibrium that is perfectly physiological. Even with every degree of freedom released, except the flexion angle, the numerical simulation for the flexion angles of 0° and 40° has a very good concordance with the measured experimental data. For higher flexion angles the results were promising. The FE simulation for the flexion angle of 55° was still close to the experimental data, with most of its nodes within 4 mm of their experimental counterpart. And for the flexion angle of 70° we can notice that only one movement has to be taken care of: an anterior translation.

This model will have to be completed in order to achieve a better stability, mostly for the higher flexion angles. We are currently working on the addition of the menisci in this model. We hope that this will allow us to obtain a lower friction coefficient and address to the anterior translation in the numerical simulation of the 70° flexion angle. After that the next step would be the modelling of the ligaments as the 3D parts with contacts that they are, and check their influence on our current results. As long as possible we will keep focusing on a passive stability-oriented model, in order to control each of its aspect.

References

Donahue TLH et al (2002) A finite element model of the human knee joint for the study of tibio-femoral contact. J Biomech Eng 124(3):273–280

Dufour M, Pillu M (2007) Biomécanique fonctionnelle: membres, tête, tronc. Rappels anatomiques, stabilité, mobilités, contraintes

Gardiner JC, Weiss JA (2003) Subject-specific finite element analysis of the human medial collateral ligament during valgus knee loading. J Orthop Res 21(6):1098–1106

Kaufman KR et al (2001) Gait characteristics of patients with knee osteoarthritis. J Biomech 34(7):907–915

Marouane H, Shirazi-Adl A, Hashemi J (2015) Quantification of the role of tibial posterior slope in knee joint mechanics and ACL force in simulated gait. J Biomech 48(10):1899–1905

Mow VC, Ateshian GA (1997) Lubrication and wear of diarthrodial joints. Basic Orthop Biomech 2:275–315

Oh KJ, Park WM, Kim K, Kim YH (2014) Quantification of soft tissue balance in total knee arthroplasty using finite element analysis. Comput Methods Biomech Biomed Eng 17(14):1630–1634

Peat G, McCarney R, Croft P (2001) Knee pain and osteoarthritis in older adults: a review of community burden and current use of primary health care. Ann Rheum Dis 60(2):91–97

Pena E et al (2006) A three-dimensional finite element analysis of the combined behavior of ligaments and menisci in the healthy human knee joint. J Biomech 39(9):1686–1701

Shandiz MA, Boulos P, Saevarsson SK, Yoo S, Miller S, Anglin C (2016) Changes in knee kinematics following total knee arthroplasty. Proc Inst Mech Eng Part H J Eng Med 230: 265–278

Wybier M, Bossard P (2013) Musculoskeletal imaging in progress: the EOS imaging system. Joint Bone Spine 80(3):238–243

Yao J et al (2006) Stresses and strains in the medial meniscus of an ACL deficient knee under anterior loading: a finite element analysis with image-based experimental validation. J Biomech Eng 128(1):135–141

Zhang C et al (2016) Robust iterative closest point algorithm with bounded rotation angle for 2D registration. Neurocomputing 195:172–180

Zhong Y et al (2011) Stress changes of lateral collateral ligament at different knee flexion with or without displaced movements: a 3-dimensional finite element analysis. Chin J Traumatol 14(2):79–83

Preclinical Analysis to Assess Aseptic Loosening of Orthopaedic Implants

Heidi-Lynn Ploeg, Anthony G. Au, Ameet K. Aiyangar, Nipun Yamdagni, Kristopher K. Biegler, Matthew W. Squire and Richard L. Illgen II

Abstract Although it is long accepted that aseptic loosening is the main reason for revision of total joint replacements, preclinical assessment methods of primary fixation are not well developed. Reasons for aseptic loosening are multifactorial including the patient, surgical approach, biological reactions, wear, micro-motion at the implant–bone interface, load transfer from the joint to the host bone, and bone adaptation. The objective of this study was to highlight a few preclinical methods to investigate orthopaedic implant primary fixation relative to the transfer of loads and displacements at the implant–bone interface. The last generation of metal-on-metal hip prostheses used a high-precision low clearance bearing to provide a low friction ball and socket joint. During implantation the acetabular component deforms under a press-fit; however, excessive deformation of the acetabular component can lead to premature failure of the joint replacement. It is therefore important to establish an accurate method of quantifying cup deformation and develop finite element models to better understand the effects of the press-fit. Methods to measure press-fit deformation of monoblock acetabular cups for metal-on-metal total hip arthroplasty and resurfacing were investigated. The purpose of the present study was to compare

H.-L. Ploeg (✉) · A.G. Au · N. Yamdagni · K.K. Biegler
Department of Mechanical Engineering, University of Wisconsin-Madison,
3047 Mechanical Engineering Building, Madison, WI 53706, USA
e-mail: ploeg@engr.wisc.edu

H.-L. Ploeg · M.W. Squire · R.L. Illgen II
Department of Orthopedics and Rehabilitation, University of Wisconsin-Madison,
Madison, WI, USA

H.-L. Ploeg
Department of Biomedical Engineering, University of Wisconsin-Madison,
Madison, WI, USA

A.K. Aiyangar
Department of Mechanical Systems Engineering, Empa,
Swiss Federal Laboratories for Materials Science and Technology,
Duebendorf, Switzerland

A.K. Aiyangar
Department of Orthopaedic Surgery, University of Pittsburgh, Pittsburgh, PA, USA

A. Gefen and D. Weihs (eds.), *Computer Methods in Biomechanics and Biomedical Engineering*, Lecture Notes in Bioengineering,
DOI 10.1007/978-3-319-59764-5_15

cup deformation with two experiments simulating press-fit of an acetabular cup into the pelvis. Rim deformation and cup strain were measured for two common tests: (1) a two-point pinching of the cup rim and (2) a press-fit implantation into a cavitated polyurethane foam block. In the pinch test, the rim displaced linearly and symmetrically with force. The press-fit test, ostensibly a closer representation of surgical procedure, produced more complex displacement and strain responses due to the foam block shape, and the cup surface-foam block interaction. The current study demonstrated two methods to measure real-time hip cup deformation and strain during press-fitting that may be used for preclinical assessment of primary fixation.

Introduction

Although it is long accepted that aseptic loosening is the main reason for revision of total joint replacements (Garellick et al. 2013; Hip and Knee Arthroplasty 2015; Sadoghi et al. 2013), preclinical assessment methods of primary fixation are not well developed. Reasons for aseptic loosening are multifactorial including the patient, surgical approach, biological reactions, wear, micro-motion at the implant–bone interface, load transfer from the joint to the host bone, and bone adaptation. The objective of this study was to highlight different preclinical methods to investigate orthopaedic implant primary fixation relative to the transfer of loads and displacements at the implant–bone interface.

Metal-on-Metal (MoM) articulations for total hip arthroplasty (THA) and hip resurfacing were developed as an alternative bearing surface to ultra-high molecular weight polyethylene (UHMWPE), with many MoM bearing couples showing promising short- to medium-term results (McMinn et al. 1996; De Smet 2005). Most MoM hip resurfacing and THA prosthesis designs employ a cobalt chromium (Co–Cr) alloy monoblock acetabular component with a highly polished, spherical articular bearing surface to facilitate low wear rates (Rieker and Kottig 2002; Jin et al. 2006). The high strength Co–Cr alloy enables thinner acetabular components and larger diameter MoM bearing couples compared to conventional metal-on-UHMWPE articulations (Clarke et al. 2005), allowing increased hip range of motion and reduced probability of hip dislocation (Jin et al. 2006; Liu et al. 2005).

MoM monoblock acetabular components are implanted with an interference press-fit between the component and an under-reamed acetabulum. The surgeon prepares a hemispherical acetabular bone bed with an inner diameter 1–2 mm less than the outer diameter of the acetabular prosthesis, and then forcefully impacts the prosthesis into the bone bed (Widmer et al. 2002; Squire et al. 2006). A successful hip surgery should include (1) correct radial clearance (i.e., the difference between outer femoral head radius and inner acetabular bearing radius) to reduce wear (Liu et al. 2006) and (2) proper cup fixation in the acetabulum to promote osseointegration (Jin et al. 2006). Acetabular cups often include a roughened outer surface and fins to improve fixation to the underlying bone. Despite these design features some cups have had unacceptably early reoperation rates (Illgen et al. 2010; Long et al. 2010). Poor understanding of the mechanical behaviour of the implanted acetabular cup (e.g. deformation, fixation) (Long et al. 2010; Fristche et al. 2008) hinders identification and prevention of the underlying causes of such failures. One possible solution is to comprehensively assess performance indicators such as radial clearance and fixation stability at the preclinical stage.

Deformation at the cup rim is key to determining whether sufficient clearance is present between the cup and femoral head after press-fit insertion (i.e. radial clearance). Also, cup stiffness is a common measure to compare the mechanical behaviour of cup designs. Cup deformation is commonly measured using a coordinate measuring machine (CMM). CMM can adequately measure deformation along the entire cup surface; however, due to its design and inherent low data acquisition rate, the measurements cannot be carried out real-time during cup insertion; and therefore are taken post-insertion. Analysing real-time cup deformation during the entire press-fit insertion process provides more insight into cup performance (Fristche et al. 2008). CMM cannot capture the real-time deformation changes occurring during insertion, and interrupting the loading mid-insertion for measurement may result in unrepresentative deformations due to relaxation of the supporting bone. The amount of relaxation has not been quantified for press-fit acetabular cups but it may be similar to relaxation levels measured in press-fit femoral knee components by Burgers et al. (Burgers et al. 2010).

The purpose of the present study was to demonstrate methods to measure real-time cup deformation and compare the mechanical behaviour depending on test method. Deformations were characterized from a combination of rim displacements and inner cup wall strains. Displacement transducers were developed specifically to measure instantaneous rim displacement. Strain gauged cups were also investigated to test their ability to estimate rim displacements.

Methods and Materials

Measurement System

Two displacement transducers were custom-made to measure diametral cup displacement along two mutually perpendicular axes at the rim of the cup. The displacement transducers were made out of 15 mm wide, aluminium alloy strips bent into a semi-circular shape with two strain gauges (Vishay Micro-Measurements, Malvern, PA) adhered to the top and bottom, respectively, of each transducer and connected in a full Wheatstone bridge network.

Four fixtures were used to secure the ends of each transducer to the cup. Each fixture was glued to the inside wall of the cup using epoxy and the transducer was secured to the fixture with a screw. One transducer ("Axial") was used to measure the displacement along the direction of the pinch/press-fit force while the second transducer ("Transverse") was used to measure the transverse displacement (Fig. 1).

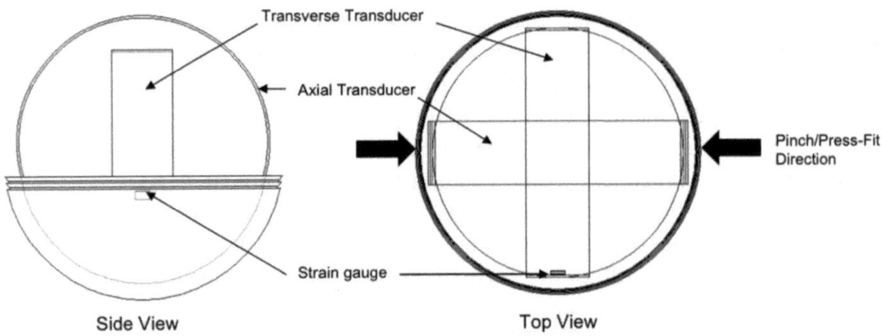

Fig. 1 Schematic showing displacement transducers and strain gauge configuration in the acetabular cup. The Axial displacement transducer measured the displacement along the direction of the pinch/press-fit force. The transverse displacement transducer measured the displacement transverse to the direction of the pinch/press-fit force. The strain gauge was oriented to measure strains parallel to the rim

A uniaxial strain gauge (CEA-06-125UN-350, Vishay Micro-Measurements, Malvern, PA), with a narrow gauge area and 3.18 mm gauge length, was placed on the inner surface of the cup 5 mm below the rim at the transverse displacement (Fig. 1). The strain gauge was oriented to measure strains parallel to the rim. To reduce the effect of temperature change, a gauge with a thermal expansion coefficient of 0.6 was chosen because it was the closest to that of Co–Cr, i.e. 0.64. To further reduce the effects of temperature, all mounting and tests were performed at room temperature, approximately 24 °C. To aid in heat dissipation, a 350 Ω strain gauge was used.

The strain gauge and displacement transducers were connected to an integrated strain gauge conditioner and amplifier system (2100 System, Instruments Division, Measurements Group, Inc., Raleigh, NC). Implantation force, transducer displacements and cup strains were simultaneously recorded at 20 Hz with an in-house program written using LabView (National Instruments Corporation, Austin, TX).

Displacement Transducer Calibration

The displacement transducers were calibrated using a sliding-stage micrometer with a resolution of 5 μm (UMR 12.40, Newport, Irvine, CA) at two ranges 0.0–0.2 mm and 1.0–1.2 mm. Within each range, the displacement transducers were first compressed and then stretched. Linear regression curves were fit to the displacement-voltage data for each displacement transducer. The procedure was repeated three times resulting in 12 curves with coefficients of determination (r^2) ranging from 0.995 to 1.000 for the Axial displacement transducer and 0.997 to 0.999 for the Transverse transducer. The calibration equations shown below were determined by averaging the 12 slopes. Displacements obtained using the transducers were accurate to within 1.8% for a range of 0–1200 μm. The transducers displaced up to 290 μm in the tests, which were well within the calibrated range of the transducers.

$$\text{Displacement} = \text{Voltage} * 10.093\,(\text{Axial transducer})$$
$$\text{Displacement} = \text{Voltage} * 6.599\,(\text{Transverse transducer}),$$

where *Displacement* is in mm and *Voltage* is in V.

Monoblock Metal-on-Metal Acetabular Cup

The Durom™ cup (size 66; 66 mm diameter reamer; inner cup diameter: 60 mm; outer cup diameter: 68 mm; Zimmer Inc., Warsaw, IN) was used as an example monoblock MoM acetabular cup (Fig. 2). It was not the intent of this study to analyse the efficacy of this particular design. The cup subtended at an angle of 165 degrees and had a constant wall thickness of 3.7 mm. The outer surface of the cup had a 0.3 mm thick vacuum plasma sprayed titanium coating. Fixation of the cup was primarily accomplished with a 2 mm diametrical press-fit and enhanced by lateral fins at the outer edge of the cup.

Test Configurations

Pinch Test

The acetabular cup was deformed under two common benchtop mechanical tests simulating press-fit. The first was a "pinch" test in which a compressive displacement was applied transversely to the rim of the cup using a universal testing

Fig. 2 The Durom™ acetabular cup (Zimmer Inc., Warsaw, IN)

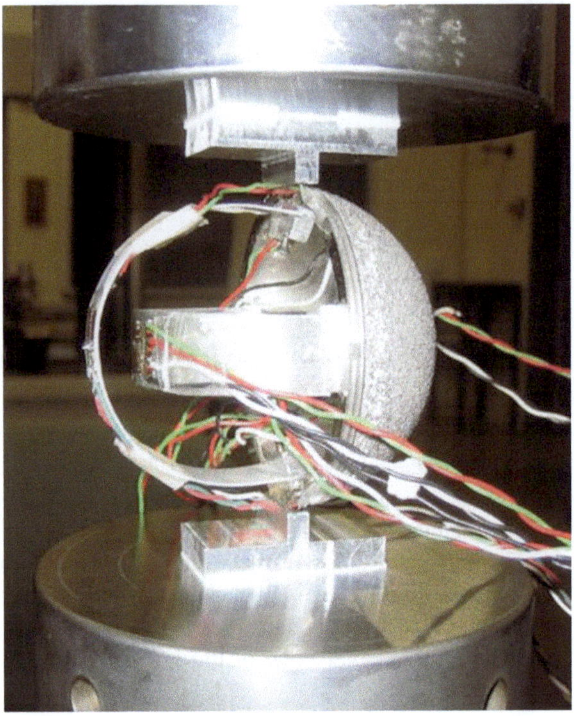

Fig. 3 Instrumented acetabular cup in preparation for pinch test. Compression was applied to the cup through custom-designed load platens

machine (MTS Sintech 10GL, Eden Prairie, MN) with a quasi-static displacement rate of 5 mm/min (Fig. 3). Rim loading was used to simulate the tight peripheral rim fit produced by pelvic compression of the cortical bone along an axis that runs from the anterosuperior to the posteroinferior margin of the acetabulum (Squire et al. 2006; Hothan et al. 2011). Loading was applied through custom-designed load platens mated to the fin geometry of the cup (Fig. 3). Pinch force, displacement transducer and strain gauge measurements were recorded at 10 N intervals between 0 and 200 N. Tests with the same cup were repeated six times.

Press-Fit Test

Acetabular cup press-fit has been simulated previously using a foam block with either a full hemispherical cavity to simulate a 360° circumferential interference fit (Fristche et al. 2008) or a cavity modified with a relief to simulate cup compression between pelvic ischial and ilial columns (Jin et al. 2006). The present study employed a relief block since maximum diametral cup deformations in the relief block ($\sim$30 μm) are more similar to human cadaveric pelves ($\sim$20 μm) than non-relief blocks (Jin et al. 2006). The acetabulum was simulated by a cavity with two areas relieved on opposite sides, compressing the cup along a single axis, similar to the model used by Jin et al. (2006) (Fig. 4).

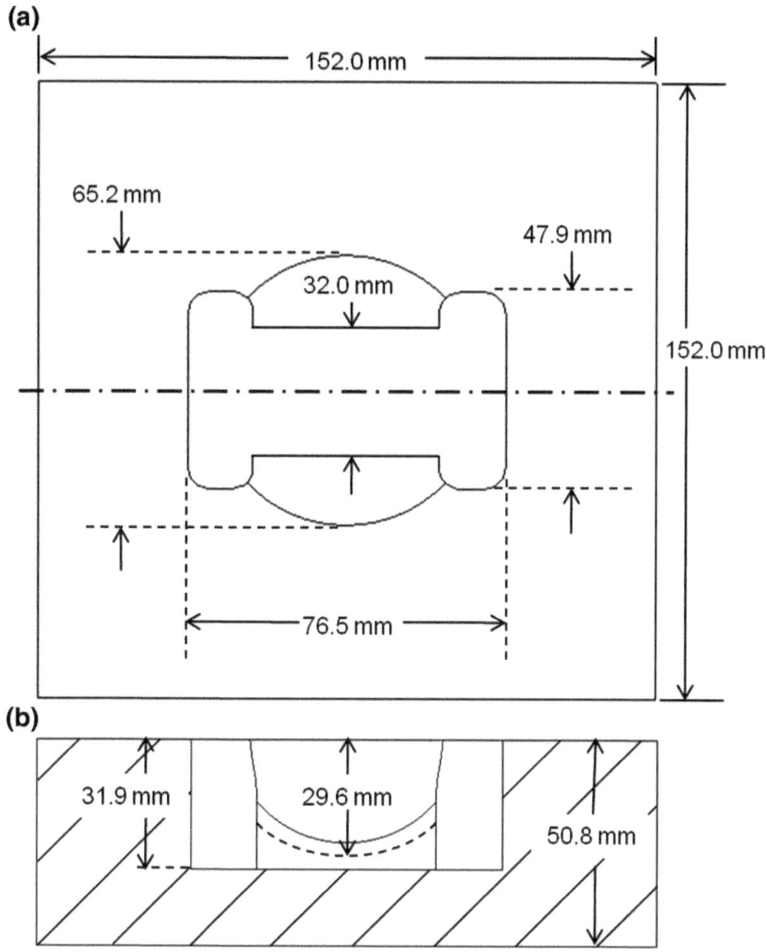

Fig. 4 Cavity dimensions of foam block employed in press-fit test seen from **a** above and **b** through a central cross-section. *Dotted arc* in (**b**) shows the border of the cavity before the 32 mm wide section at the polar region of the cavity was removed

A computer numerically controlled (CNC) mill (Haas Automation, Oxnard, CA) was used to prepare the cavity in a 50.8 mm (2-in.) thick solid rigid polyurethane foam block with a density of 48 lb/ft^3 and a compressive Young's modulus of 423 MPa (McMaster-Carr, Elmhurst, IL). The cavity was shaped to provide a 2 mm diametral interference press-fit predominantly at the rim of the cup. A 32 mm wide section at the polar region of the spherical cavity was removed to prevent contact at the pole of the cup. A 1 mm × 45° chamfer was machined into the circular rim of the cavity to reduce the 1 mm diametral interference from the fins. The cup was press-fit into the foam block using the above testing machine at a displacement rate of 5 mm/min (Fig. 5).

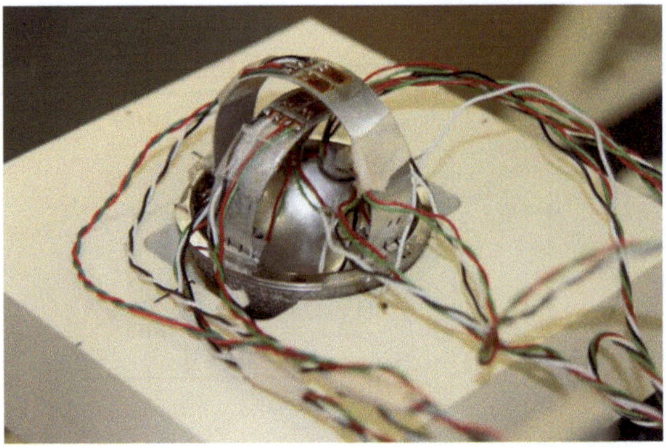

Fig. 5 Instrumented acetabular cup seated in foam block in preparation for press-fit test

A custom-made loading fixture was used to accommodate the protruding transducers and its attached wires. The fixture was machined out of a mild steel tube, with an inside diameter of 64 mm and an outside diameter of 69 mm, similar to the inner and outer cup diameters. Three archways were cut into the bottom of the loading fixture to accommodate the transducer wires. At the beginning of the test, a small preload (<10 N) was applied to the cup to ensure level seating in the cavity. Implantation force, crosshead displacement, cup strain and transducer displacement data were acquired simultaneously at a sampling rate of 20 Hz using LabView (National Instruments, Austin, TX). Tests were repeated in seven different foam blocks with the same cup.

Data Analysis

The linearity of the rim displacement and strain were assessed using a first-order regression analysis. Rim displacement versus strain plots were analysed to determine if inner wall strains could be a surrogate measure for rim displacements. The cup stiffness was determined from the slope of the force-displacement curve obtained with the pinch test. Deformation symmetry was determined by comparing axial and transverse rim displacements. A two-tailed unpaired *Student's t test* was used to determine if rim displacements in the axial and transverse directions were significantly different with a 95% significance level.

Results and Discussion

The clinical failures associated with recent cup designs highlight the need to fully assess cup performance before clinical implementation. Preclinical testing can provide valuable information regarding implant performance, but the proper performance measures must be assessed. Standardized benchtop mechanical tests (e.g. ISO, ASTM) are used preclinically to assess suitability of many joint implant designs, but none currently exist for evaluating the fixation of hip resurfacing cups. The pinch and press-fit benchtop mechanical tests suggested for simulating press-fit (Jin et al. 2006; Squire et al. 2006) were employed in this study. Instruments enabling instantaneous displacement and strain measurements were used in each test to observe deformation trends that occurred during the press-fit insertion.

Pinch Test

The pinch test mechanically simulated the pelvic compression of an acetabular cup during THA by compressing the cup uniaxially between platens of a load frame (Squire et al. 2006). The simple procedure provides a convenient method for estimating cup stiffness from design characteristics such as changing wall thickness (Squire et al. 2006) and the effect of inlay insertion across different cup manufacturers and models (Hothan et al. 2011). The computed stiffness of the cup used in this study was 1154 N/mm.

The simple cup deformation from the pinch test provided a good platform for preliminary testing of the instantaneous measurement system. The 200 N pinch force compressed the cup rim up to 178 ± 12 μm along the loading direction and expanded it up to 173 ± 11 μm in the transverse direction. Rim displacement magnitudes in the two directions were statistically identical throughout loading, indicating that the cup deformed symmetrically (i.e. slope in Fig. 6 not significantly

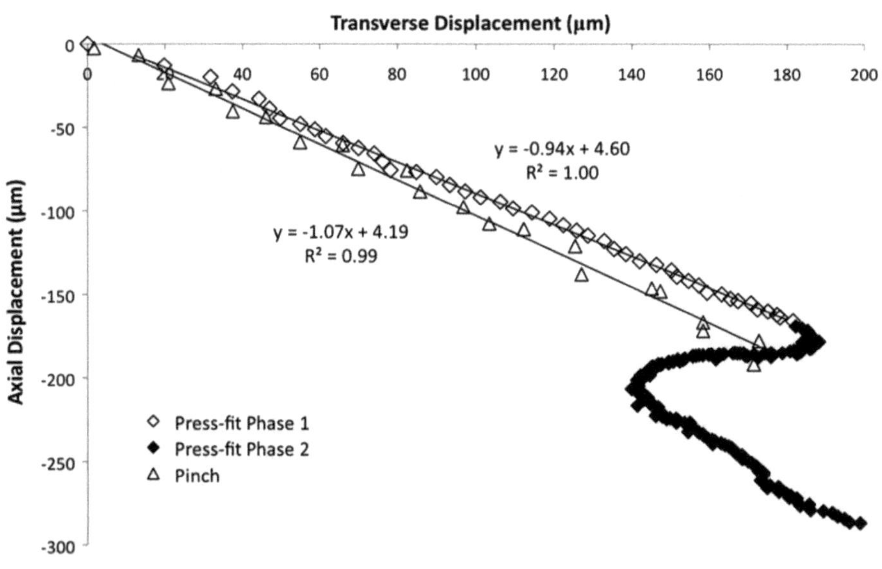

Fig. 6 Relationship between transverse and axial displacements for pinch and press-fit tests

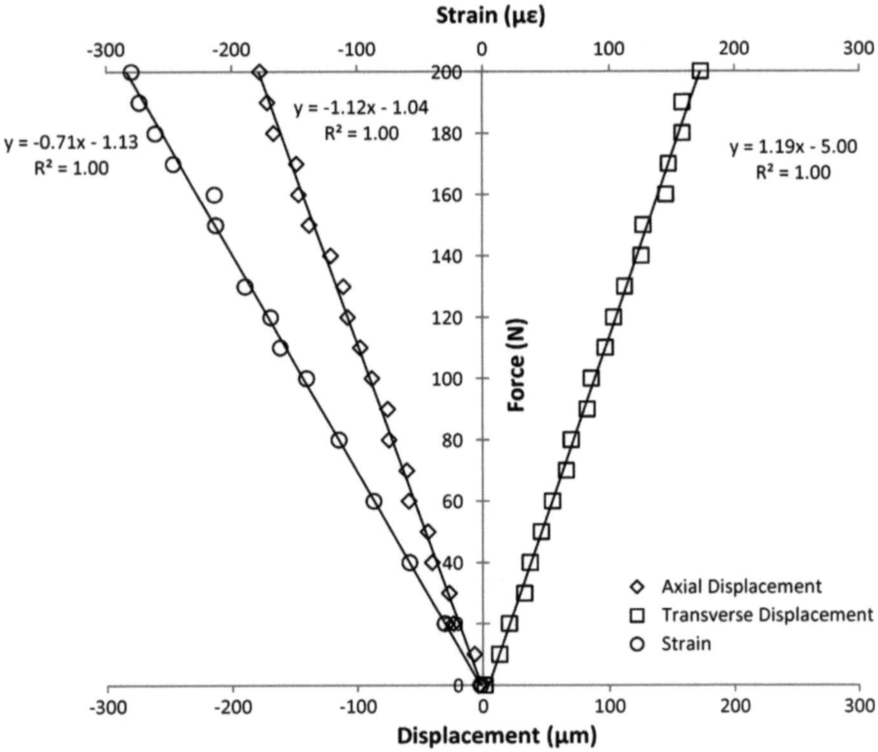

Fig. 7 Force-displacement and force-strain curves for the pinch test. Linear regression analysis equations are also shown

different from 1, $p = 0.395$). Rim displacements and pinch force were linearly correlated (Fig. 7), similar to the trend reported by Squire et al. (2006). The linearity suggests that, when a pinch test is employed, mid-insertion displacements can be confidently predicted from pre- and post-insertion displacements measured from CMM or callipers. Strains were also strongly correlated with rim displacements (Fig. 8). Calibrated strain gauges can therefore be used to predict rim displacement when it is inconvenient to use transducers.

Press-Fit Test

The press-fit test simulates the pelvic acetabulum boundary conditions better than the pinch test by using a modified polyurethane foam block that (1) supports the entire cup and (2) enables press-fit interference using an under-reamed cavity (Jin et al. 2006). This provides a more sophisticated platform for preclinical assessment of acetabular cup performance that accounts for the influence of fixation features, such as threads and fins (Fristche et al. 2008). Fixation of the fins elicited a

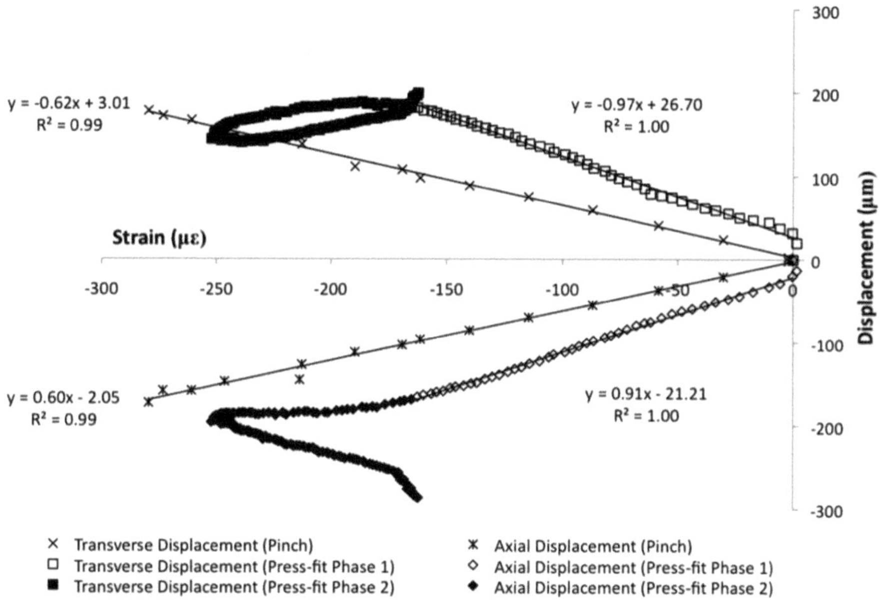

Fig. 8 Relationship between cup strain and rim displacements for pinch and press-fit tests

more complex deformation pattern, which was not revealed in the pinch test. In particular, deformation during the press-fit displayed both linear and nonlinear behaviours.

Prior to fin engagement (referred to as Phase 1), the rim displaced in the press-fit (Axial) and transverse directions in a linear fashion with crosshead displacement (Fig. 9). The displacements in the two directions were not significantly different from each other (slope = 0.941, $p = 0.697$) (Fig. 6), indicating a symmetric deformation pattern. After fin engagement (i.e., Phase 2), however, rim displacements varied nonlinearly with crosshead displacement, alternating between expansion and compression (Fig. 9). When the cup was fully seated, the rim had an asymmetric shape: axial displacement magnitude (290 $\pm$ 31 μm) was significantly greater than transverse displacement magnitude (202 $\pm$ 59 μm) ($p = 0.007$). The cup likely did not plastically deform since the maximum strain experienced by the cup (181 με) was an order of magnitude below the yield strain of the Co–Cr alloy (ISO 5832-12 2007).

It was hypothesized that the fins were the main cause of the asymmetric rim displacement. As the fins contacted the block and penetrated it, the foam fractured locally around the fins and released some of the resistance (force) offered by the block, resulting in temporary rim expansion. Once the fins fully engaged the block, the axial rim compression resumed. Fin engagement may have also caused irregular/insufficient fixation of the lower portion of cup in the latter implantation stages, as suggested by the strain measurements. Before fin engagement, the strains showed that the cup was in compression 90° from the pinch direction, 5 mm below the rim (Figs. 2, 9). However, once the fins engaged, the compressive strain

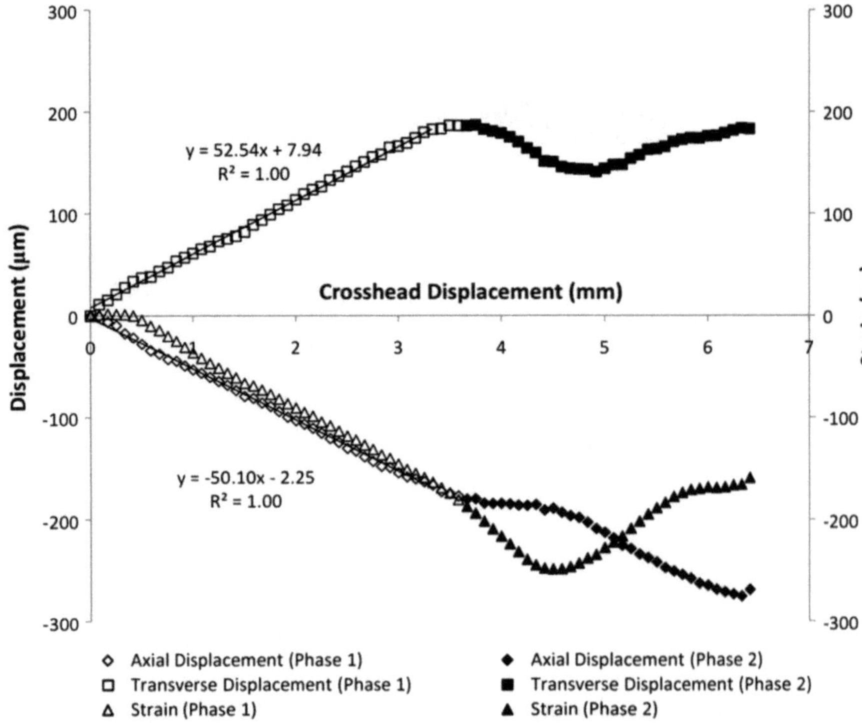

Fig. 9 The relationships of rim displacements and strain with crosshead displacement for press-fit test. Linear regression analysis equations between strain and rim displacements with crosshead displacement for Phase 1 of press-fit test are shown. No strain appeared for the initial 0.4 mm of crosshead displacement because the polar region of the cup had not yet contacted the foam block

increased with crosshead displacement. The relaxation of the strains suggests that, once the fins engaged, fixation of the cup was diverted away from the cup surface to the fins. This may have resulted in the cup not being fully seated in the cavity, causing either a gap or weak contact between the outer cup surface and the inner surface of the cavity. Physiologically, a weak contact interface or gap between the cup and surrounding acetabular bone retards/prevents osseointegration and denies an important avenue of secondary fixation. If the cup behaviour seen in this study can be confirmed to occur postoperatively, it may explain the lack of bone ingrowth on the implant surface and subsequent loosening of the acetabular cups seen clinically by Long et al. (2010) and Illgen et al. (2010).

Strain as Surrogate Measure for Rim Displacement

Strain gauges provided localized and direction-dependent measurements of the inner cup surface. Strain data must be interpreted with caution if they are used to estimate rim displacement. The relationship between strains and displacements is not intuitive and a calibration must be performed even for simple boundary conditions. Ostensibly, displacement and strain may be related via cup diameter based on mechanics of materials theory (Fristche et al. 2008). In the pinch test, displacement and strain shared a strong relationship but were not related through cup diameter (Fig. 8). Using this relationship, the strain gauge can be used to estimate rim displacement when it is inconvenient to use the transducers. Moreover, in the press-fit test, it was not possible to accurately predict rim displacement from strain after the lateral fins engaged the foam.

Limitations

The polyurethane foam in this study had similar elastic modulus (423 MPa) to the Sawbones® (Vashon Island, WA) grade 30 foam (445 MPa) employed in previous press-fit studies (Jin et al. 2006; Fristche et al. 2008), but several clues suggest the present foam block was too stiff and unrepresentative of human pelvic bone. First, the fully seated cup rim displaced up to 290 μm. This was higher than reported in similar press-fit studies, but within the calibrated range of the transducers. This could be due partly to different cup design, which is known to affect insertion behaviour (Fristche et al. 2008) and cup deformation (Jin et al. 2006). Second, the cup employed in this study had two levels of lateral fins progressively closer to the rim. After testing and removal of the cup, one set of grooves was observed in the foam suggesting that only the lower level of fins cut into the foam. The top level of fins pressed against the top surface of the foam block but never engaged the block despite the extremely high implantation force (42.47 $\pm$ 0.24 kN) applied to the cup. Finally, prior to fin engagement, the cup had a tendency to eject out of the cavity, i.e., bounce back.

It was not the intention of this study to assess the clinical accuracy of the benchtop simulations or the performance of the acetabular cup. It should be acknowledged, however, that both the pinch and press-fit tests employed simplified boundary conditions compared to the in vivo situation. Widmer et al. (2002) showed that a press-fit acetabular cup under simulated one-legged stance was primarily loaded at three locations: the iliac bone, the ischial facet, and the pubic bone. The pinch test simplified this by using a two-point contact at the rim and ignored the bone interface at the lower cup, while the press-fit test assumed the cup was loaded at the ilial and ischial columns. A follow-up study will use the measurement system to observe press-fit deformation trends in human cadaveric pelves to provide data that may be more closely representative of the intraoperative condition.

The surface of the cavity was not measured in this study but it has been shown that even machined cavities can be slightly non-spherical and moderately affect cup deformation (Jin et al. 2006). This resulted in small variances in the strains and rim displacements of the cup during press-fit insertion into the foam block.

Summary and Conclusions

Unacceptably high early revision rates of recent MoM THA cup designs highlight the need to fully assess mechanical behaviour of new designs before clinical implementation. This study demonstrated the importance of analysing cup deformation throughout the entire insertion process, as it can provide important clues regarding the extent of interface contact. Rim displacement and inner cup wall strain trends were measured real-time during two press-fit simulations: a two-point pinching of the cup and press-fit implantation into a foam block.

In the pinch test, rim displacements and strains varied linearly with pinch force; and, the cup deformed in a symmetrical manner. The press-fit test, in contrast, produced a complex asymmetrical deformation pattern that was influenced by the lateral fins. Specifically, the displacement and strain trends became nonlinear after the fins engaged the foam block. This may signal poor fixation of the lower cup due to the lateral fins. The value of such deformation and strain measurements underlies the importance of measuring deformation using instruments with real-time data acquisition capability.

Acknowledgements The authors acknowledge the Natural Sciences and Engineering Research Council of Canada (NSERC) and the University of Wisconsin Graduate School for their financial support.

References

Burgers TA, Mason J, Squire M, Ploeg H-L (2010) Time dependent fixation and implant forces for a femoral knee component: an in vitro study. Med Eng Phys 32:968–973

Clarke IC, Donaldson T, Bowsher JG, Nasser S, Takahashi T (2005) Current concepts of metal-on-metal hip resurfacing. Orthop Clin N Am 36:143–162

De Smet KA (2005) Belgium experience with metal-on-metal surface arthroplasty. Orthop Clin N Am 36:203–213

Fristche A, Bialek K, Mittelmeier W, Simnacher M, Fethke K, Wree A, Bader R (2008) Experimental investigations of the insertion and deformation behavior of press-fit and threaded acetabular cups for total hip replacement. J Orthop Sci 13:240–247

Garellick G, Kärrholm J, Lindahl H, Malchau H, Rogmark C, Rolfson O (2013) Swedish Hip Arthroplasty Register, Annual Report 2013. http://www.shpr.se/Libraries/Documents/AnnualReport_2013-04-1_1.sflb.ashx

Hip and Knee Arthroplasty (2015) Annual report 2015, Australian Orthopaedic Association National Joint Replacement Registry. OAO, Adelaide. https://aoanjrr.sahmri.com/documents/10180/217745/Hip%20and%20Knee%20Arthroplasty

Hothan A, Huber G, Weiss C, Hoffmann N, Morlock M (2011) Deformation characteristics and eigenfrequencies of press-fit acetabular cups. Clin Biomech 26:46–51

Illgen RL, Squire MW, Heiner JP, Conrad D (2010) Large head metal-on-metal total hip arthroplasty using the Durom acetabular component at minimum 1 year interval. J Arthroplasty 25:e3

ISO 5832-12 (2007) Implants for surgery—metallic materials—part 12: wrought cobalt-chromium-molybdenum alloy

Jin ZM, Meakins S, Morlock MM, Parsons P, Hardaker C, Flett M, Isaac G (2006) Deformation of press-fitted metallic resurfacing cups. Part 1: experimental simulation. Proc IMechE Part H J Eng Med 220:311–319

Liu F, Udofia IJ, Jin ZM, Hirt F, Rieker C, Roberts P, Grigoris P (2005) Comparison of contact mechanics between a total hip replacement and a hip resurfacing with a metal-on-metal articulation. Proc IMechE Part C J Mech Eng Sci 219:727–732

Liu F, Jin Z, Roberts P, Grigoris P (2006) Importance of head diameter, clearance, and cup wall thickness in elastohydrodynamic lubrication analysis of metal-on-metal hip resurfacing prostheses. J Eng Med 220:695–704

Long WT, Dastane M, Harris MJ, Wan Z, Dorr LD (2010) Failure of the Durom Metasul acetabular component. Clin Orthop Relat Res 468:400–405

McMinn D, Treacy R, Lin K, Pynsent P (1996) Metal on metal surface replacement of the hip. Clin Orthop Relat Res 329S:S89–S98

Rieker CB, Kottig P (2002) In vivo tribological performance of 231 metal-on-metal hip articulations. Hip Int 12(2):73–78

Sadoghi P, Liebensteiner M, Agreiter M, Leithner A, Bőhler N, Labek G (2013) Revision surgery after total joint arthroplasty: a complication-based analysis using worldwide arthroplasty registers. J Arthroplasty 28:1329–1332

Squire M, Griffin WL, Mason JB, Peindl RD, Odum S (2006) Acetabular component deformation with press-fit fixation. J Arthroplasty 21(Suppl 2):72–77

Widmer KH, Zurfluh B, Morscher EW (2002) Load transfer and fixation mode of press-fit acetabular sockets. J Arthroplasty 17:926–935

Study of Unloading at Lower Limbs Orthotics: Experiment and Simulation

Il'ya N. Dashevskiy, Sergey E. Nikitin, Mikhail N. Perel'muter and Pavel S. Shushpannikov

Abstract The rate of healing and rehabilitation in fractures of the feet depends on the level and the program of loading the damaged segment. The paper studied the possibilities of efficiently controlling the unloading level of the affected area by varying the degree of orthosis tightening (lateral compression). Among others, a hypothesis was tested, according to which only in a slip the limb by increasing the lateral compression is pushed up and the coefficient of unloading (CU) grows. Two types of shin-orthosis system models were developed and studied numerically for different variants of parameters: conical models using the package, based on boundary integral equation method and BEM, and personalized real form models based on the technology «CT scan –> Specialized processing software (Mimics) –> FEA package». The unloading coefficient dependence on the circumferential tightness for two types of "slippery" stocking materials was measured. When using both synthetic and silk stockings, experimental results did not show valid growth of CU with an increase in tightening (lateral compression) neither while standing nor walking. As calculations showed, pushing the shin up due to lateral compression may be to a maximum of $\sim$ 1.2 mm even in total slip. At the same time, the shift of the shin surface as a result of the piston effect can reach several mm. Thus, the additional sediment due to the so-called "piston effect" (skin stretching under a load, as well as its shift-slip over thin and soft layer of subcutaneous fat) is several times greater than the effect of ejection due to lateral compression and associated unloading, and thus completely reduces them to nothing. Further possibilities and approaches to the problem of load programming at limb orthotics are discussed.

I.N. Dashevskiy (✉) · M.N. Perel'muter · P.S. Shushpannikov
Institute for Problems in Mechanics of the Russian Academy of Sciences, Prosp. Vernadskogo 101, Block 1, Moscow, Russia 119526
e-mail: dash@ipmnet.ru

S.E. Nikitin
Federal State Unitary Enterprise "CITO" of the Ministry of Health of Russia, Moscow, Russia

A. Gefen and D. Weihs (eds.), *Computer Methods in Biomechanics and Biomedical Engineering*, Lecture Notes in Bioengineering,
DOI 10.1007/978-3-319-59764-5_16

Introduction

The rate of healing and rehabilitation in fractures of the feet depends on the level and the program of loading the damaged segment (Logvenkov and Stein 2006). In Dashevskiy and Nikitin (2016) the idea has been put forward of possibilities of efficiently controlling the unloading level of the affected area by varying the degree of orthosis tightening (lateral compression). However, experiments conducted in Dashevskiy and Nikitin (2016), Dashevskiy and Nikitin (2014), Dashevskiy (2016) did not reveal reliable dependence of the coefficient of unloading (CU) on tightening. There was also formulated a hypothesis on the causes of such an effect, according to which only in a slip the limb by increasing the lateral compression is pushed up and the CU will grow. In reality, conditions of contact between leg and orthosis are rather dry friction conditions, which depending on different parameters can lead to the formation of areas of adhesion, sliding, and detachment of different sizes. Thus, for theoretical analysis of the possibility of unloading control it is necessary to study the models of "Foot-Orthosis" system taking into account friction.

Materials and Methods

Experimental Determination of Friction Coefficients Between Skin and Stocking, Stocking and Orthosis

According to the existing procedure when applying orthosis at first stocking is put on the leg for hygienic reasons and only then the orthosis is imposed. Thus, the patient's skin contacts the stocking which external side is in contact with the orthosis sleeve. In present study, tests quite similar to those described in Dashevskiy and Nikitin (2014), Dashevskiy (2016) were conducted for stockings made of silk—the material, on the one side, natural and hygienic, and on the other side – slippery enough (Fig. 1a, b).

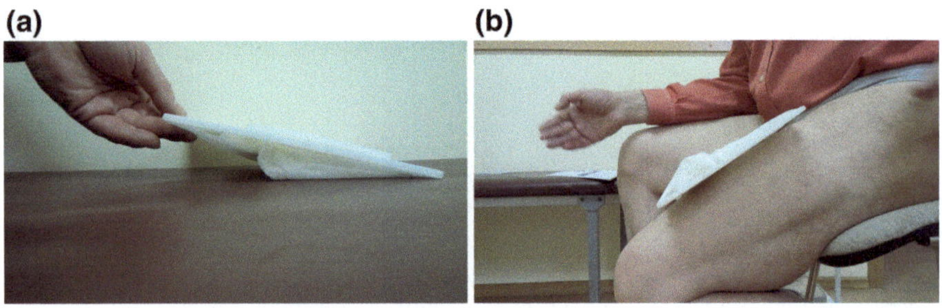

Fig. 1 Measurement of friction coefficients (the large specimen is covered with silk) between: **a** orthosis and stocking (a small plate of HDPE is placed topside); **b** the stocking and the skin

Table 1 Basic mechanical and geometrical parameters

	E (Pa)	ν	L (m)	$r_i \mid r_e \parallel r_l \parallel\mid r_u$ (m)	t (m)
b (bone)	2×10^{10}	0.35	0.185	0.001 | 0.002	–
m (muscle tissue)	27×10^3	0.5	0.185	‖ 0.035 ‖| 0.060	–
o (orthosis)	5.5×10^8	0.3	0.185	"•"	0.004

E is the Young's modulus, ν—Poisson's ratio, L—the length of the shin-orthosis contact zone, r_i r_e, r_l, r_u—respectively, the inner and outer radii of the bone, and the radii of the shin at its narrowest (lower) and widest (upper) ends, and t—thickness of the orthosis wall. Body weight $P = 87$ kg was simulated by application to the upper end of the bone the stresses $p = 900$ kPa (Fig. 4b), giving a total of 87 kg

Modeling: Conical Model

Two types of shin-orthosis system models were developed and studied numerically.

In conical models a hollow cylinder (tubular bone) was engaged with truncated cone (muscles), relying upon shin-conformal thin-walled cone (orthosis). At the bottom the bone rests on the arch of the foot, modeled by a spring with known stiffness $F_f = 0.15 \cdot 10^6$ Pa (Dashevskiy and Nikitin 2016) (Fig. 4a).

Table 1 summarizes the main calculation parameters (see also Fig. 4a).

Calculations used the package, based on boundary integral equation method and BEM.

Modeling: Model of Real Form

Another (personalized) approach to the problem was based on multispiral computer tomography (MSCT) of the shin according to the technology «CT scan → Specialized processing software (Mimics) → FEA package», which gives real exact description of the geometry of the limb (as well as of the orthosis) (Fig. 6). The shear modulus for muscle tissue was taken the same as previously.

Experimental Study of the Unloading Coefficient Dependence on the Circumferential Tightness for Synthetic and Silk Stockings

Measurement of feet pressure on the support was carried out with the help of software and hardware complex (SHC) "DiaSled" according to the methodology, presented in Dashevskiy and Nikitin (2016), Dashevskiy and Nikitin (2014), Dashevskiy (2016). Tests were conducted on healthy trial subject with intact limbs and individual orthosis (Fig. 2a, b).

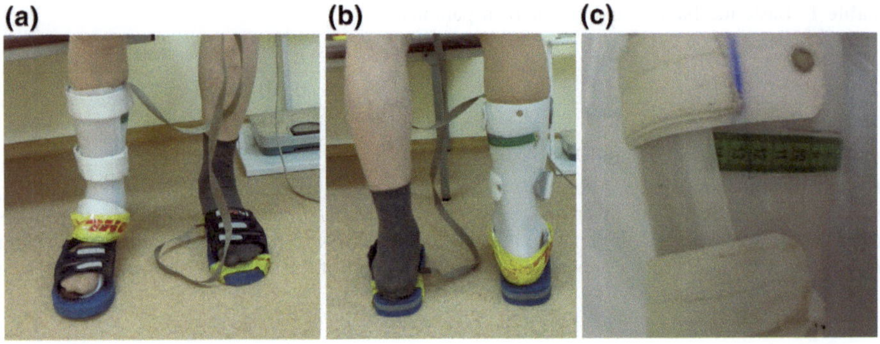

Fig. 2 **a** Trial subject in the orthosis with sensors installed of the "DiaSled" system (front view); **b** the same (back view); **c** measuring tape

Changing circumferential compression was performed by tightening the top fastening. Tightening was measured using millimeter tape pasted on the orthosis sleeve below the fastening (Fig. 2c). Figure 3 presents a typical view of SHC DiaSled windows: Fig. 3a—upon standing on each leg alternately, Fig. 3b—upon standing and when walking. The coefficient of unloading (CU) was calculated (while standing or all over the single measurement when walking) as the ratio of the maximum load on the unorthosed limb to the maximum load on the limb in an orthosis.

Results

The measured coefficients of friction for silk with HDPE (orthosis) and skin turned out to be 0.21 and 0.69, respectively.

Table 2 summarizes the results of calculations.

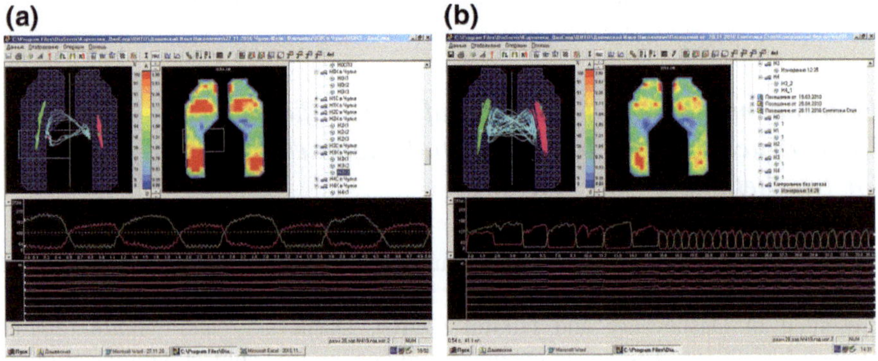

Fig. 3 Typical view of SHC DiaSled windows: **a** upon standing on each leg alternately; **b** upon standing and when walking

Table 2 Coefficient of unloading as calculated for different variants of parameters

Variant	*P*87E27f016	*f*048	5*E*	*P*/2	$f_c = 0.21$
w (mm)	5.69	5.40	4.68	5.04	5.66
p_m (kPa)	47	89	200	42	51
K	0.05 = 5%	0.1 = 10%	0.22 = 22%	0.05 = 5%	0.06 = 6%
k	1.05	1.1	1.3	1.05	1.06

*P87E27f*016 denotes the basic variant: $P = 87$ kg, $E = 27$ kPa, $f = 0.16$ (synthetic stocking), the title of each further column indicates its difference from the basic variant: $f = 0.48$ (cotton stocking), etc.; p_m—part of the load transferred on the orthosis case, $K = p_m/p$, $k = 1/(1–K)$—coefficient of unloading

Figures 4 and 5 illustrate the main results obtained.

Figure 6 shows the distribution of areas of seizing (red), sliding (green), and detachment (blue) with $k = 0.48$ (left, cotton) and $k = 0.16$ (right, synthetics). It can be seen that according to the calculations at $k = 0.16$ there are no areas of seizing at all,

The plots of CU on the tightness built on the data obtained, and have the form as shown in Fig. 7.

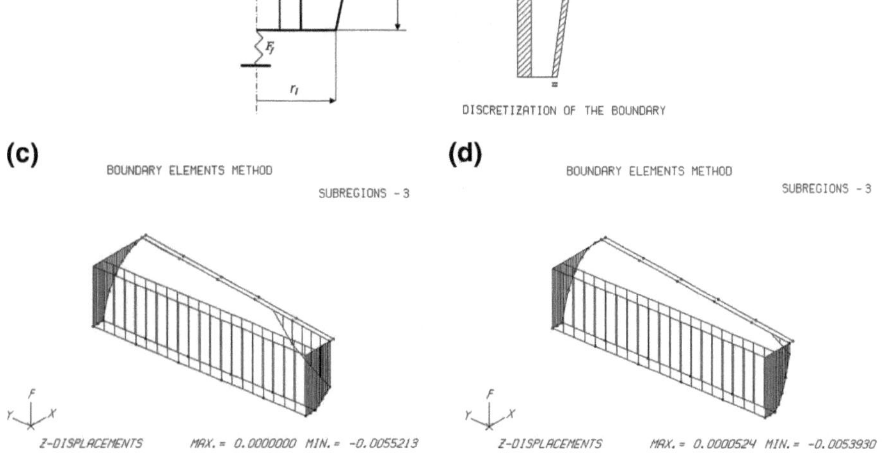

Fig. 4 **a** Conical model of a shin-orthosis system; r_b, r_l, r_u, t_l, t_u, *L*—characteristic dimensions, *P*—body weight, being transmitted to the tubular bone; **b** discretization of the boundaries; **c** The basic variant of calculations; **g** cotton stocking: $f = 0.48$, $w = 5.40$ mm, $p_m = 89$ kPa

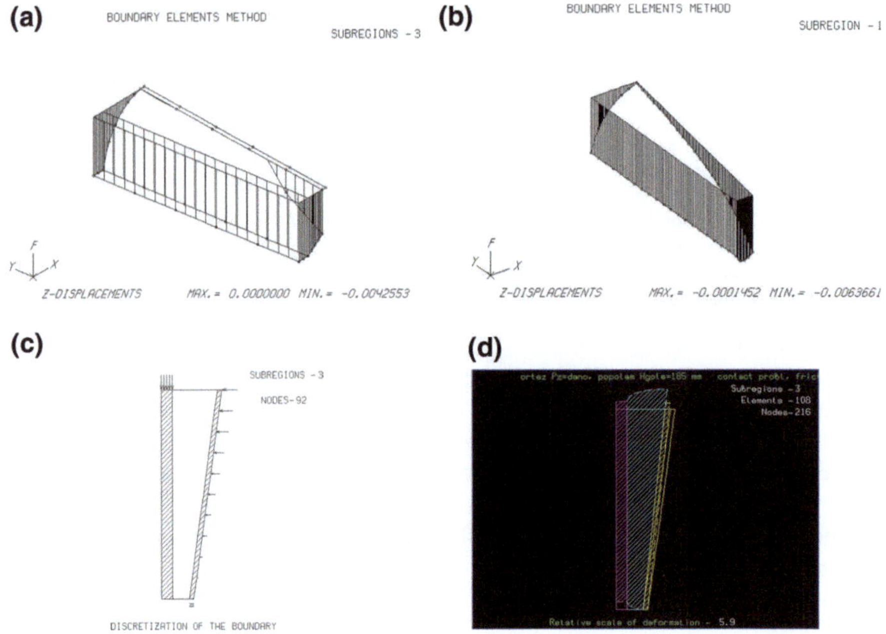

Fig. 5 **a** The effect of shin elastic modules: $E \rightarrow E' = 5E$; **b** displacement diagrams of the muscular layer boundaries for $f = f_c = 0.21$, when at the first time there is no one point of the shin adhesion to the orthosis; **c** modeling of the lateral tightening prescribing displacements at the boundary; **d** deformations at maximum circumferential tightening $\Delta l = 0.04$ m

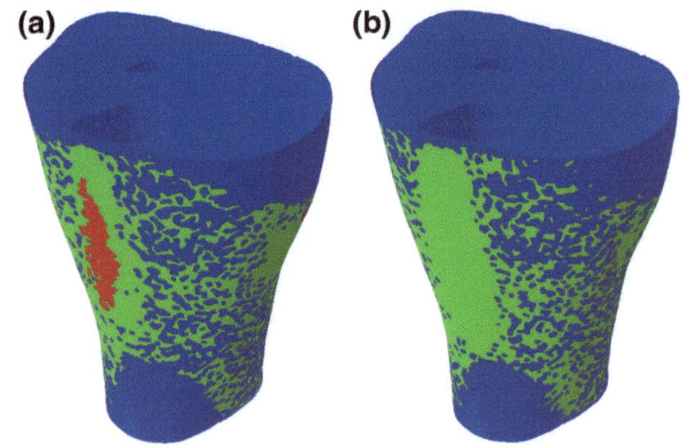

Fig. 6 Arrangement of areas of adhesion, sliding, and the lack of contact on the Shin–Orthosis interface for **a** $k = 0.48$ (cotton); **b** $k = 0.16$ (synthetics)

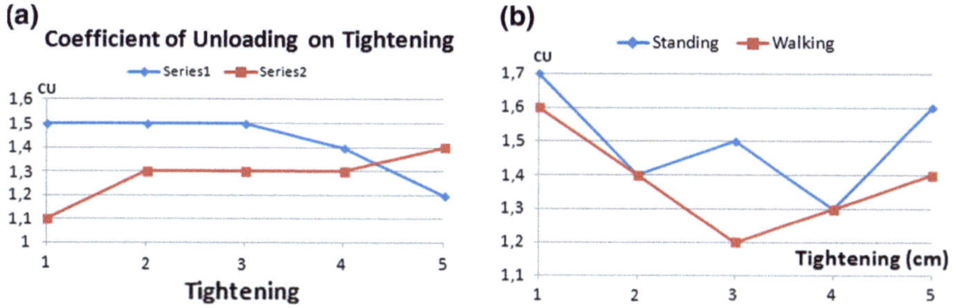

Fig. 7 Plots of unloading coefficient on the tightness: **a** for synthetic stockings; **b** for silk stockings

Discussion

As shown from Fig. 7a, b when using both synthetic and silk stockings experimental results do not show valid growth of CU with an increase in tightening (lateral compression) neither while standing nor walking. For both types of stockings CU values were higher while standing than when walking.

We tend to believe that such a result (no compression effect on unloading) is due to the so-called "piston effect": (skin stretching under a load, as well as its shift slip over thin and soft layer of subcutaneous fat). Indeed, according to Dashevskiy and Nikitin (2016) measurements gave for the stiffness of the foot arch the value $F_f = 0.15 \cdot 10^6$ Pa. With a weight of $P = 87$ kg this gives a subsidence $w = 5{,}684 \approx 5.7$ mm. As calculations showed, pushing the shin up due to lateral compression may be to a maximum of ~ 1.2 mm. At the same time, as is easily seen directly, the shift of the shin surface as a result of the piston effect can reach several mm. Thus, the additional sediment due to the piston effect is several times greater than the effect of ejection due to lateral compression and associated unloading, and thus completely reduces them to nothing.

There might be other approaches to the problem of load programming at limb orthotics. As already been indicated (Dashevskiy and Nikitin 2016), this approach may be associated with such positioning and fixing limbs, which would provide a low initial load on the segment. As the recovery proceeds the load can be gradually increased, easing the tightness. The limiting case here is the creation of the technical gap between the foot and the orthosis sole when applying orthosis (e.g., using an appropriate gasket inserted between the foot and the orthosis when taking the cast of the limb). More complex schemes may include adjustable gaps created with special devices (runners and screws, etc.), spring, or other paddywhack tabs which would allow both to regulate and, coming together with the strain gauges in the sensor unit, to control the load on limb (Aydarov 2010).

However, the main thing is controlling the level of loading a limb. To do this, apparently some simple sensor may be used placed between the heel and the orthosis and serving for orthosis individual customization to provide adequate load on the affected segment of the limb.

Acknowledgements This work was supported by RFBR 14-08-01266 and 14-08-00537 Grants.

References

Aydarov VI (2010) Orthopedic device "foot holder" as used in the postoperative period after corrective shin bone osteotomy. In: RAS innovations–2010: materials of the annual scientific conference. Publishing house "Word", Kazan, Russia, pp 193–194 (in Russian)

Dashevskiy IN (2016) On possibility of unloading control at leg orthotics. Proc Struct Integr 2:1277–1284

Dashevskiy IN, Nikitin SE (2014) Development of the experimental method and measurement of unloading and other characteristics of the shin-orthosis system. Preprint No. 1091, RAS Institute for Problems in Mechanics, Moscow (in Russian)

Dashevskiy IN, Nikitin SE (2016) Biomechanics of unloading of the lower extremities at orthotics. Russ J Biomech 20(2):113–127

Logvenkov SA, Stein AA (2006) Control of biological growth as a problem of mechanics. Russ J Biomech 10(2):9–18

Statistical Shape Model Based 2D–3D Reconstruction of the Proximal Femur—Influence of Radiographic Femoral Orientation on Reconstruction Accuracy

J. O'Connor, M. Rutherford, J. Hill, D. Beverland, N. Dunne and A. Lennon

Abstract In recent years, it has been suggested that statistical shape model based 2D–3D reconstruction of the proximal femur could offer a solution to issues with preoperative planning of total hip replacement from 2D radiographs. The purpose of this work was to assess the effect of radiographic femoral rotation on the accuracy of a statistical shape model based 2D–3D femoral reconstruction method. A reconstruction algorithm was tested on input images with varying amounts of internal/external rotation (10 internal to 50 external) using leave-one-out tests and the resulting 3D shape was compared to the CT segmentation by point-to-point distance. The minimum value for mean point-to-point error was 1.24 ± 0.18 mm and occurred at 20° of external rotation (where neutral orientation was the femoral neck axis aligned in the coronal plane). The maximum error calculated was at 50° of external rotation with mean point-to-point error being 1.88 ± 0.41 mm. This work highlights an important source of error for 2D–3D reconstruction algorithms which may be incorporated into future validation studies.

J. O'Connor (✉) · M. Rutherford · A. Lennon
School of Mechanical and Aerospace Engineering, Queen's University Belfast, Belfast, UK
e-mail: joconnor15@qub.ac.uk

J. Hill · D. Beverland
Outcomes Assessment Unit, Musgrave Park Hospital, Stockman's Lane, Belfast, UK

N. Dunne
School of Pharmacy, Queen's University Belfast, Belfast, UK

N. Dunne
School of Mechanical and Manufacturing Engineering, Dublin City University, Dublin, Ireland

N. Dunne
Trinity Centre for Bioengineering, Trinity College Dublin (TCD), Dublin, Ireland

A. Gefen and D. Weihs (eds.), *Computer Methods in Biomechanics and Biomedical Engineering*, Lecture Notes in Bioengineering,
DOI 10.1007/978-3-319-59764-5_17

Table 1 2D/Reconstruction of the femur highlighting overall accuracy based on a defined validation method

Author	Anatomical region	2D–3D method	Validation method	Error
Le Bras et al. (2003)	Proximal femur	Non-stereo corresponding contours method	25 cadaveric femora, biplane low-dose EOS™ images	0.7 mm point-to-surface distance (P2S)
Baudoin et al. (2008)	Proximal femur	Parametric subject specific model	13 cadaveric femora, biplane low-dose EOS™ images	1.0 mm P2S
Zheng et al. (2009)	Proximal femur	Point distribution model and regularised shape deformation	22 cadaveric femora biplane X-ray images	0.9 mm mean distance
Yu et al. (2016)	Proximal femur	Control point based deformation of a 3D volume template	10 cadaveric femora + 44 biplane digitally reconstructed radiographs (DRRs)	1.29 mm for cadaveric and 0.9 mm for DRRs

Introduction

Preoperative planning and postoperative assessment of Total Hip Replacement (THR) are achieved predominantly using two-dimensional (2D) X-ray imaging despite indications that femoral orientation during image acquisition may lead to unreliable measurements (Lechler et al. 2014). Three-dimensional (3D) planning based on computed tomography (CT) scans of the hip have shown improved prediction of implant selection and placement when compared to conventional 2D planning (Sariali et al. 2012). However, several factors inhibit the implementation of routine CT scanning for THR patients, including the additional cost and increased radiation dose relative to traditional X-ray imaging. Recent advances in the 3D femoral reconstruction using radiographs (Table 1) have shown potential to provide 3D shape for preoperative and postoperative THR assessment, the issue being that their application in surgical planning has shown poor prediction of implant size (Zheng et al. 2016).

Validation methods for these reconstructions use dry cadaveric femora or CT scans to generate digitally reconstructed radiographs (DRRs) followed by reconstruction based on the 2D images and comparison to the original 3D ground truth shape. The input 2D images are in a known 3D orientation despite considerable variation in the clinical environment—i.e. there tends to be an unknown degree of femoral rotation on preoperative radiographs (Lechler et al. 2014). The purpose of this work is to assess the effect of radiographic femoral rotation on the accuracy of a statistical shape model (SSM) based 2D–3D reconstruction method.

Materials and Methods

Statistical Shape Model Building

A sample of 3D femora ($n = 47$) were segmented from the Virtual Skeleton Database (SICAS, Swiss Institute for Computer Assisted Surgery, Switzerland) (Kistler et al. 2013). A reference femur was chosen and the remaining femora were rigidly registered to it using a Procrustes analysis in Shape Modelling software Scalismo (R0.12, Graphics and Vision Research Group, University of Basel, Switzerland) (Lüthi et al. 2013; Clogenson et al. 2014). Point-to-point correspondences were established by fitting a Gaussian process model of the reference shape to every femur in the database. Principal component analysis was then performed to identify modes of variation and their associated variance (Lüthi et al. 2013; Clogenson et al. 2014).

2D/3D Reconstruction

A custom algorithm was developed in MATLAB (R2015b, The MathWorks, Inc., MA, USA) for fitting the SSM to femoral contours, it was assumed that contours were extracted from the X-ray and the input to the program is two sets of points, each representing femoral contours extracted from an antero-posterior (AP) and lateral (LT) radiograph. Along with deformation based on the SSM, the algorithm allowed internal/external (IE) rotation (i.e. rotation around the femoral axis) so that each shape and orientation can be described by Eq. 1 below.

$$x = \begin{bmatrix} \mathrm{Cos}(\theta) & -\mathrm{Sin}(\theta) & 0 \\ \mathrm{Sin}(\theta) & \mathrm{Cos}(\theta) & 0 \\ 0 & 0 & 1 \end{bmatrix} \left(\bar{x} + \sum_{i=1}^{m} \alpha_i \boldsymbol{P}_i \right), \tag{1}$$

where $\boldsymbol{x}$ is a 3D candidate shape with IE rotation (θ), $\bar{\boldsymbol{x}}$ is the mean shape, $\boldsymbol{P}_i$ are the eigenvectors of the covariance matrix which define the possible shape deformations, $\boldsymbol{\alpha}_i$ are the shape parameters and $\boldsymbol{m}$ is the number of modes of variation used. Each generated shape is transformed to the neutral orientation with the *Z*-axis parallel to the femoral shaft axis, *Y*-axis parallel a line between the femoral head center and shaft center and the *X*-axis orthogonal to *Z*- and *Y*-axes.

$$\boldsymbol{SA}_p = \beta_A \boldsymbol{SA} \quad \beta_A \in \mathbb{R},\ \beta_A \geq 1 \tag{2}$$

Comparison between candidate solutions and input contours is achieved in 2D by projection of the 3D solution using a ray tracing algorithm (Eq. 2) and step 2 in Fig. 1 (Freud et al. 2006). Where $\boldsymbol{A_p}$ were the coordinates of the projection of point $\boldsymbol{A}$ and $\boldsymbol{\beta_A}$ is the distance from the source ($\boldsymbol{S}$) to the point divided by the distance

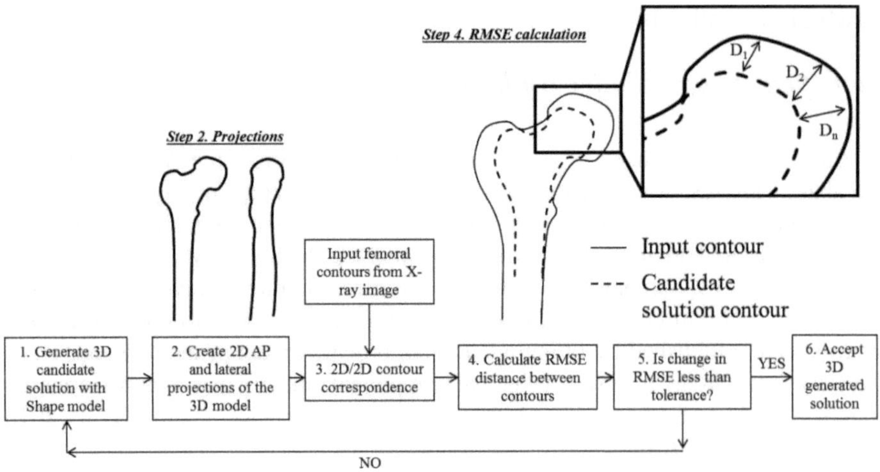

Fig. 1 Overview of 2D/3D reconstruction algorithm for fitting statistical shape model to input contours

from the source to the detector. The source to detector distance was 1.2 m and the beam was centred 150 mm medial of the femoral head to represent the clinical case were the beam is centred on the pubic symphysis. This method provided the 2D projection for every point in the reference mesh, to reduce subsequent computation; the built-in boundary function in MATLAB was used to identify only the outer surface points of the projection. In order to assess the error between the possible solution and the target contours, correspondence needed to be established between the contours. The Coherent Point Drift algorithm was used for rigid registration and 2D/2D correspondence within the loop allowing for an error metric to be calculated (Myronenko and Song 2009). The objective function used here was the Root Mean Squared Error (RMSE) between projection of the candidate solution and the input contours. Since there were two contours used for optimisation, the sum of the RMSE values for each contour was used to guide the solution. A built-in single-variable function minimiser was used in MATLAB to solve sequentially for the rotation and shape parameters that best matched the input contours. Rotation was solved using the mean shape projection and shape parameters were solved independently with each solution brought forward to the iteration for the next mode (i.e. Mode 1 solution is stored and carried to the solver for Mode 2 solution).

Algorithm Testing

Leave-one-out tests were performed by rebuilding the SSM (n = 46) without the femur to be tested, creating AP and LT projections of the left out femur (ground truth) and using the 2D/3D algorithm to reconstruct the femoral surface based on the projections. The solution was then rigidly registered to the ground truth femoral surface using an Iterative Closest Point algorithm, correspondence was calculated using nearest neighbour interpolation. Point-to-point distance could then be evaluated to determine the distance from ground truth to the 8720 points in the solution mesh.

Influence of Radiographic Rotation on Reconstruction Accuracy

The influence of radiographic femoral rotation on reconstruction accuracy was tested using the method described in Sect. 2.3 with the input femoral contours being generated at specified angles of IE rotation. Since the database contained 47 femora and one must be left out for each test, 46 femora were available for testing.

Rotations ranging from 10° internal to 50° external in increments of 10° (Fig. 2) were tested and results were presented as boxplot distributions of point-to-point distance.

Results

Reconstruction Accuracy and Speed

When the input contours were generated with the femur in neutral orientation, the mean distance between corresponding points was 1.38 ± 0.35 mm for all femora. The most accurate reconstruction had a mean error of 0.82 ± 0.44 mm and the least accurate an average point-to-point error of 2.25 ± 1.45 mm (Fig. 3). The reconstruction program produced a surface model in approximately 3 min on a standard desktop with a 3.40 GHz Intel quad core i7-2600 processor and 16 GB RAM.

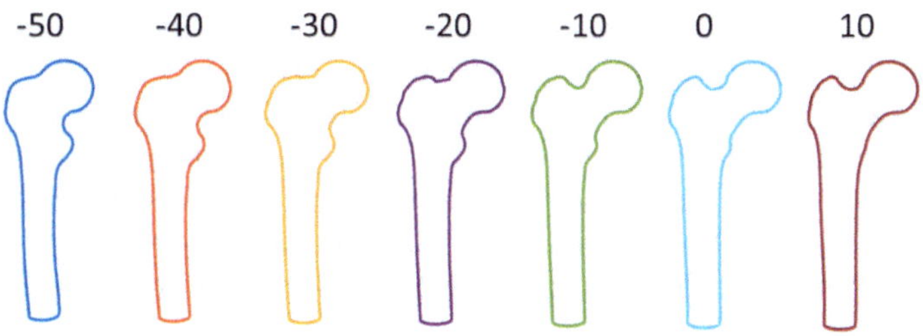

Fig. 2 Projected contours resulting from 3D IE rotation of the femur from 50° External (−) to 10° Internal (+)

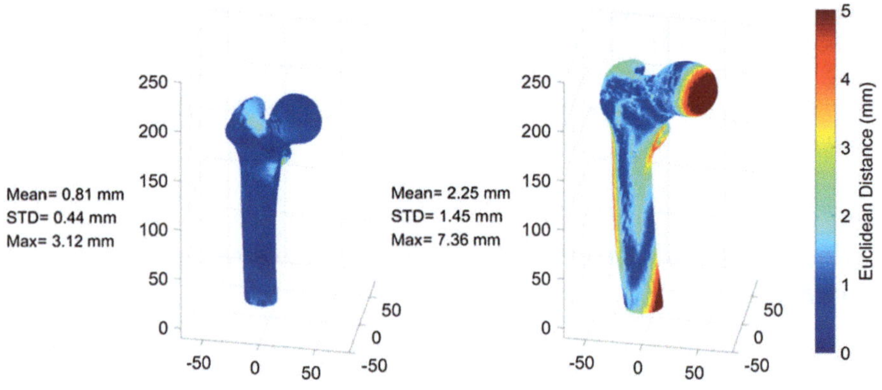

Fig. 3 Femur with the most accurate reconstruction (*left*) and femur with the least accurate (*right*)

Fig. 4 Reconstruction accuracy (point-to-point distance) with input contours generated at specified IE rotation

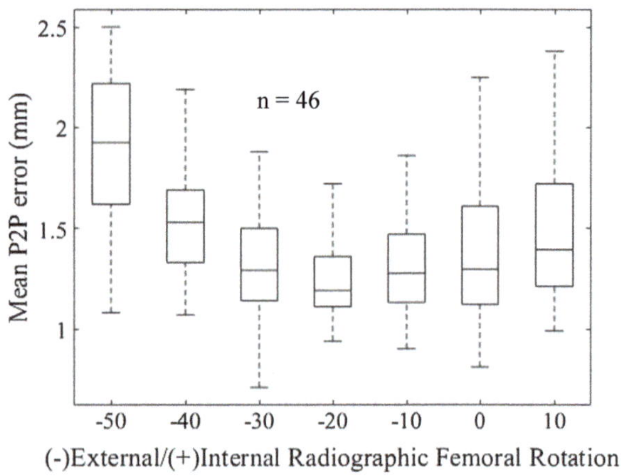

Influence of Rotation of Accuracy

The minimum value for mean P2P error was 1.24 ± 0.18 mm and occurred at 20° of external rotation. The maximum error calculated was at 50° of external rotation with mean P2P error being 1.88 ± 0.41 mm (Fig. 4).

A Kruskal–Wallis test was performed in MATLAB and p values less than 0.01 were considered statistically significant. The only group with a significantly higher mean when compared to the neutrally orientated group was the 50° externally rotated group ($p < 0.001$). Reconstruction error from a contour with 20° of external rotation was statistically lower than 40° and 50° external groups as well as the 10° internal rotation group ($p < 0.01$).

Discussion

Reconstruction accuracy at neutral (1.38 mm) and time (3 min) to find a solution compared favourably with recent literature (Table 1) albeit tested on contours without the presence of both magnification and segmentation errors. It is worth noting also that the work of Yu et al. (Table 1) is volume reconstruction and therefore more prone to error than surface reconstruction presented here (Yu et al. 2016). The addition of a rotational degree of freedom to the possible solution may reduce the accuracy of the algorithm when neutrally oriented input is assumed. Since a SSM is constrained by the variance in the training data, recent work has applied a further regularised shape deformation after SSM fitting leading to increased accuracy of the solution (Zheng et al. 2009). We have not incorporated this in our model; this reduces model flexibility to fit to shapes under-represented in the training data and may affect the solution accuracy. Although the minimum reconstruction error is presented at 20° of external rotation, there was no statistical difference when compared to neutral, 10° external or 30° external, which may suggest that the method is relatively insensitive to IE rotation within that range. Larger errors were observed at the extremes of the ranges chosen—i.e. 50° external and 10° internal, this is likely due to the loss of definition in the contour in the trochanter areas of the curve (Fig. 2). The definition for neutral in this study was the orientation that maximised radiographic femoral offset which reflects the imaging protocol for THR where the leg is internally rotated by 15° in order to reduce errors caused by femoral neck anteversion (Clohisy et al. 2008). As a result, 15° of external rotation in this study is, on average, equivalent to anatomical neutral, the orientation of the femur in an AP projection of a CT scan and dry femur setup (provided they are not corrected for femoral anteversion).

References

Baudoin A, Skalli W, de Guise JA, Mitton D (2008) Parametric subject-specific model for in vivo 3D reconstruction using bi-planar X-rays: application to the upper femoral extremity. Med Biol Eng Comput 46(8):799–805

Clogenson M, Duff JM, Luethi M et al (2014) A statistical shape model of the human second cervical vertebra. Int J Comput Assist Radiol Surg 30:1097–1107

Clohisy JC, Carlisle JC, Beaulé PE et al (2008) Young adult hip a systematic approach to the plain radiographic evaluation of the. J Bone Jt Surg Am 90:47–66

Freud N, Duvauchelle P, Létang JM, Babot D (2006) Fast and robust ray casting algorithms for virtual X-ray imaging. Nucl Instrum Methods Phys Res Sect B Beam Interact Mater Atoms 248 (1):175–180

Kistler M, Bonaretti S, Pfahrer M, Niklaus R, Büchler P (2013) The virtual skeleton database: an open access repository for biomedical research and collaboration. J Med Internet Res 15(11): e245

Le Bras A, Laporte S, Bousson V et al (2003) Personalised 3D reconstruction of proximal femur from low-dose digital biplanar radiographs. In: CARS 2003 computer assisted radiology and surgery. Proceedings of the 17th international congress and exhibition, vol 1256, pp 214–219

Lechler P, Frink M, Gulati A et al (2014) The influence of hip rotation on femoral offset in plain radiographs. Acta Orthop 85(4):389–395

Lüthi M, Jud C, Vetter T (2013) A unified approach to shape model fitting and non-rigid registration. In: International workshop on machine learning in medical imaging, no 1, pp 66–73

Myronenko A, Song X (2009) Point-set registration: coherent point drift. Adv Neural Inf Process Syst 19:1009–1016

Sariali E, Mauprivez R, Khiami F, Pascal-Mousselard H, Catonné Y (2012) Accuracy of the preoperative planning for cementless total hip arthroplasty. a randomised comparison between three-dimensional computerised planning and conventional templating. Orthop Traumatol Surg Res 98(2):151–158

Yu W, Chu C, Tannast M, Zheng G (2016) Fully automatic reconstruction of personalized 3D volumes of the proximal femur from 2D X-ray images. Int J Comput Assist Radiol Surg 11:1673–1685

Zheng G, Gollmer S, Schumann S, Dong X, Feilkas T, González Ballester MA (2009) A 2D/3D correspondence building method for reconstruction of a patient-specific 3D bone surface model using point distribution models and calibrated X-ray images. Med Image Anal 13(6):883–899

Zheng G, Schumann S, Balestra S, Thelen B, Nolte L-P (2016) 2D–3D reconstruction-based planning of total hip arthroplasty. In: Zheng G, Li S (eds) Computational radiology for orthopaedic interventions. Springer, Cham, pp 197–215

Comparison of Different Strain-Based Parameters to Identify Human Left Ventricular Myocardial Infarct During Diastole: A 3D Finite-Element Study

Gerardo Kenny Rumindo, Jacques Ohayon, Magalie Viallon, Mathias Stuber, Pierre Croisille and Patrick Clarysse

Abstract Assessment of myocardial damage is important to obtain an accurate prognosis after myocardial infarction. Myocardial strains have been shown to be good indicators of the myocardial viability. Strain analysis has been used to identify dysfunctional regions, and several strain-based parameters have been proposed to detect regions of infarct. In this study, nine different parameters were used to detect synthetically generated infarct lesions on left ventricles and their performances were compared. The parameters were investigated on ten virtual 3D cases based on healthy human left ventricles extracted from MRI examinations. Realistic infarcts were generated for each virtual case with different locations, shapes, sizes, and stiffer materials. Diastolic virtual strain data were obtained via finite-element simulations using widely implemented constitutive law and rule-based myofiber orientation. The results showed that stretch-dependent invariance in fiber direction is able to better delineate the infarcts in comparison to the other parameters.

G.K. Rumindo (✉) · M. Viallon · P. Croisille · P. Clarysse
Univ Lyon, INSA-Lyon, Université Claude Bernard Lyon 1, UJM-Saint Etienne, CNRS, Inserm, CREATIS UMR 5220, U1206, Lyon, France
e-mail: Kenny.Rumindo@creatis.insa-lyon.fr

J. Ohayon
TIMC-IMAG Laboratory, University Grenoble Alpes, CNRS, Grenoble, France

J. Ohayon
University of Savoie Mont-Blanc, Engineering School Polytech Annecy-Chambéry, Le Bourget du Lac, France

M. Stuber
Center for Biomedical Imaging and Center for Cardiovascular Magnetic Resonance Research, University of Lausanne, Lausanne, Switzerland

A. Gefen and D. Weihs (eds.), *Computer Methods in Biomechanics and Biomedical Engineering*, Lecture Notes in Bioengineering,
DOI 10.1007/978-3-319-59764-5_18

Introduction

Accurate assessment of myocardial viability is important to determine the clinical course of heart disease patients (Allman 2002; Bhat 2016). A decline in myocardial viability negatively influences the structural and functional properties of the heart, which typically leads to heart dysfunctions. Late gadolinium enhancement (LGE), a cardiac magnetic resonance (CMR) imaging technique, is currently the gold standard to identify the location of myocardial infarcts (Doltra 2013). Strain analysis has also been used to identify dysfunctional myocardial regions. Myocardial strains, resulting from cardiac motion, can be extracted via tagged CMR or tissue tracking technique (Prince and McVeigh 1992; Schuster 2015). Several strain-based parameters have been proposed to detect regions of infarct, such as principal strains (Lee 2008) and local directional strains (Goette 2001). Fractional anisotropy, a technique commonly used in diffusion tensor field, has also been investigated for the purpose of localizing myocardial infarcts (Soleimanifard 2010). Although systolic and diastolic dysfunctions co-exist in heart disease patients (Elahi 2014; Warren and Grossman 1991), most research in this area investigated the strains in the systolic phase of the cardiac cycle to detect the infarct regions (Duchateau 2016; Soleimanifard 2010).

In this work, we compared nine different strain-based parameters to detect synthetically generated infarct lesions on left ventricles (LVs) in the diastolic phase. In addition to the strain-based parameters mentioned previously, effective strain as well as stretch-dependent invariances in fiber, sheet, and sheet-normal direction (Holzapfel and Ogden 2009) were also tested. The ability of each strain-based parameter to detect the synthetic infarcts in the myocardium in the diastolic phase was evaluated.

Materials and Methods

Five healthy LV meshes, each consisting of 3552 hexahedral elements, were obtained from Genet (2014). Finite-element mechanical model of each healthy LV in the diastolic phase was generated following the inverse problem optimization as shown in Fig. 1. Next, ten pathological cases were generated from these five healthy LV models with arbitrary size, shape, and location.

Simulations of Healthy Heart Models

Following the pipeline shown in Fig. 1, a rule-based myocardial fiber orientation was assigned to each LV in accordance to the algorithm described in Genet (2015). Briefly, a pseudo-prolate spheroidal coordinate system was defined based on the

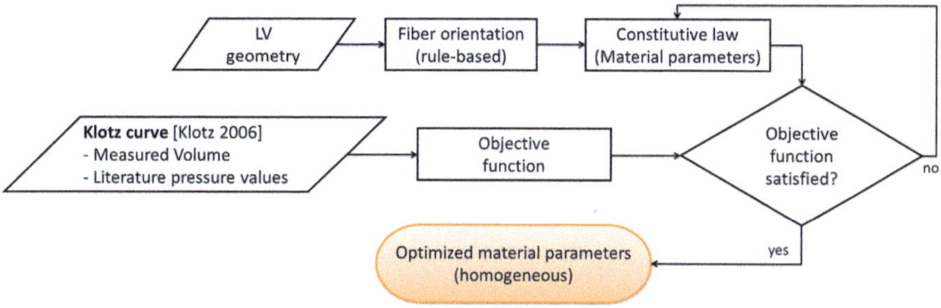

Fig. 1 Inverse optimization to obtain the personalized material parameters for the healthy LV models

$$\mathrm{W} = \frac{C}{2}(e^{\bar{Q}} - 1) + \frac{1}{D}(\frac{J^2 - 1}{2} - lnJ)$$

$$\bar{Q} = b_f E_{11}^2 + b_t E_{22}^2 + b_t E_{33}^2 + b_t(E_{23}^2 + E_{32}^2) + b_{fs}(E_{12}^2 + E_{21}^2 + E_{13}^2 + E_{31}^2)$$

Isochoric term — Volumetric term — Exponential term of the isochoric part, as a function of Green-Lagrange strain tensor ***E***

Fig. 2 Transversely isotropic Fung-type law used to simulate diastolic phase

centroid of each element. The pseudo-prolate coordinate system consists of transmural, circumferential, and longitudinal positions. The transmural position is normalized to 0 at the endocardium and 1 at the epicardium; the circumferential position is the angle around the LV axis, which was calculated from the barycenter of the geometry and the manually selected apex point; and the longitudinal position is also normalized to 0 at the apex and 1 at the basal level. Furthermore, the transmural basis vector is the distance-weighted average of the outward-pointing normal of the endocardial and epicardial surfaces closest to the element centroid; the circumferential basis vector is the circumferential vector around the ventricular axis; and the longitudinal basis vector is simply the cross product of the first two basis vectors. Finally, the basis vectors were rotated about the transmural basis vector in accordance to the varying elevation angle of +60° to −60° from the endocardium to the epicardium (Streeter 1969).

The constitutive law used to simulate the diastolic filling phase in this study is the transversely isotropic Fung-type law (Genet 2014), as shown in Fig. 2. It was implemented in ABAQUS user material subroutine (UMAT) using Fortran language (ABAQUS 2011). For implementation in ABAQUS UMAT, the strain energy density is decomposed into volumetric and isochoric terms. The volumetric term is associated with a large bulk modulus to enforce quasi-incompressibility, thereby D was set to 0.001. The isochoric part is based on Green-Lagrange strain with 1 isotropic constant C, and 3 exponential constants b_f, b_t, and b_{fs}. The variables that need to be updated for each increment in the UMAT is the Cauchy stress and the corotational tensor of elasticity, which were calculated based on the formulae

Table 1 Material parameters for the healthy LV models (Genet 2014)

Healthy model	Material parameters	
	C (kPa)	B
VA1	0.117	12.4
VA2	0.121	15.0
VA3	0.105	18.3
VA4	0.107	16.0
VA5	0.123	10.1

provided in Young (2010). For validation of the developed UMAT, a cubic specimen consisting of one linear hexahedral element with 1 integration point was subjected to uniaxial stretching and simple shear tests. The analytical solutions were compared with the ABAQUS UMAT results.

In this work, we used the optimized parameters from Genet (2014), which are listed in Table 1 for all the five healthy LV models. The ratios of b_f to b_t and b_{fs} were kept fixed to the values in Genet (2014), thereby limiting the inverse optimization problem to only two parameters; C and B, where C is the isotropic constant and B is the exponential constant where $B = b_f = 2.5\, b_t = (\frac{10}{7})\, b_{fs}$. These parameters were personalized based on the diastolic pressure–volume profile described in Klotz (2006). The volume of each subject at end-diastole and end-systole was available; whereas the end-diastolic pressure value was set at 9 mmHg. Based on this information, the optimization process is to minimize the distance error between the simulated end-diastolic pressure–volume relationships (EDPVRs) with respect to the Klotz curve.

Creation of Pathological Database and Evaluation of Strain-Based Parameters

Ten synthetic pathological cases were created from the five healthy heart models. Realistic infarcts were generated with different sizes and locations. The infarcts had truncated ellipsoidal shapes within the volume range of 3000–50,000 mm^3. In addition, the center point of the infarct geometry was defined to be in the endocardial surface. The different geometries of both the LVs and the infarct introduced a higher variability in the database in order to better investigate the sensitivity of the strain-based parameters.

The material properties of the infarcts were set to be stiffer than the healthy ones (Popovic 1996; Genet 2015), in accordance to Table 2. The deformation gradient tensors of each element, and subsequently the Green-Lagrange strain tensors, were extracted from the diastolic simulation of all pathological cases.

Nine strain-based parameters were calculated from the resulting Green-Lagrange strain tensors; first (most positive) principal strain, effective strain, fractional anisotropy, radial, circumferential, and longitudinal strains, as well as fiber, cross-fiber,

Table 2 Material and geometrical properties of the generated infarct

Infarct case	Healthy model	Material parameters	Infarct volume (mm^3)
1	VA1	C × 5	46,537
2		b_f × 4	12,056
3	VA2	All params × 1.5	13,977
4		Exp params × 2	18,510
5	VA3	Isotropic exp params	8350
6		C × 5	7918
7	VA4	All params × 2	12,555
8		Isotropic exp params	10,681
9	VA5	Exp params × 2	4763
10		b_f × 4	3579

The material of the infarcts was made stiffer than the healthy ones by multiplying the material parameters as shown in the table. C and b_f correspond to the material parameters shown in Fig. 2. '*All params*' includes C, b_f, b_t, and b_{fs}. '*Exp params*' means all the parameters excluding C. '*Isotropic exp params*' means that the values of b_f, b_t and b_{fs} were set to the same values (the value of healthy b_f), essentially removing any preferred directions

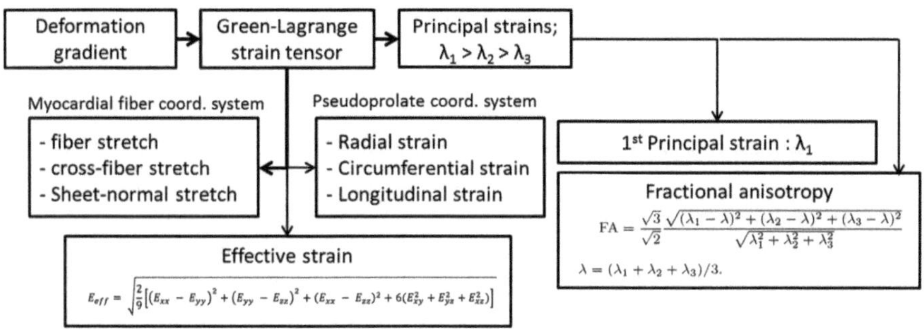

Fig. 3 The nine strain-based parameters

and sheet-normal stretches. Figure 3 shows the calculation done to obtain each strain-based parameter.

We calculated the volume-normalized scalar mean value for each strain-based parameter in the infarct region. This value was calculated by observing the parameter values in the infarct regions of the pathological cases and the corresponding regions in the healthy cases. The sum of parameter value at each element was then normalized to the total volume of the infarct region. The difference in the volume-normalized parameter values between the healthy and pathological case was considered to show some degree of sensitivity of each strain-based parameter for the respective infarct regions. We introduced the discrimination index S, calculated as the root mean square value of these differences. Thus, the sensitivity of each strain-based parameter could be represented and compared with simplified scalar indices.

Results

ABAQUS UMAT Validation

Figure 4 shows the validation results of our user-defined material implementation in ABAQUS UMAT with respect to the analytical solutions. Two uniaxial stretching and three simple shear tests were conducted to validate the developed user material.

Simulations of 5 Healthy LVs

The resulting diastolic EDPVRs with the personalized parameters for each healthy LV model are shown in Fig. 5, which were considered still within the acceptable

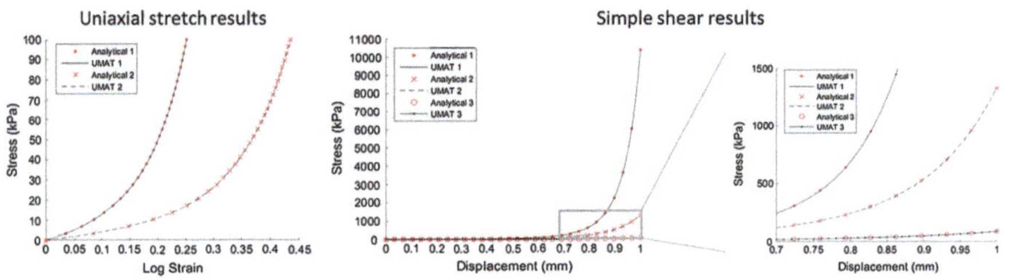

Fig. 4 Validation of the numerical (UMAT) results against the analytical solutions using 1 cubic hex element. Uniaxial stretch results (*left*): stretch in fiber (1) and cross-fiber (2) directions. Simple shear results (*right*): shear 12 (1), shear 21 (2) and shear 23 (3); shear *ij* means the face in the positive *i*-direction is sheared to the *j*-direction

Fig. 5 End-diastolic pressure–volume curves of the healthy heart models against the Klotz curve

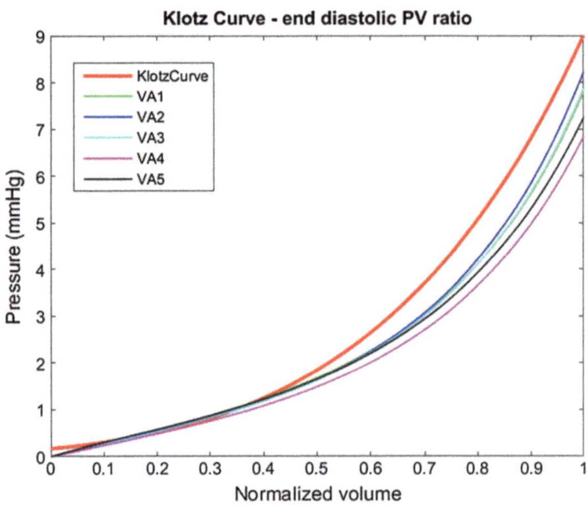

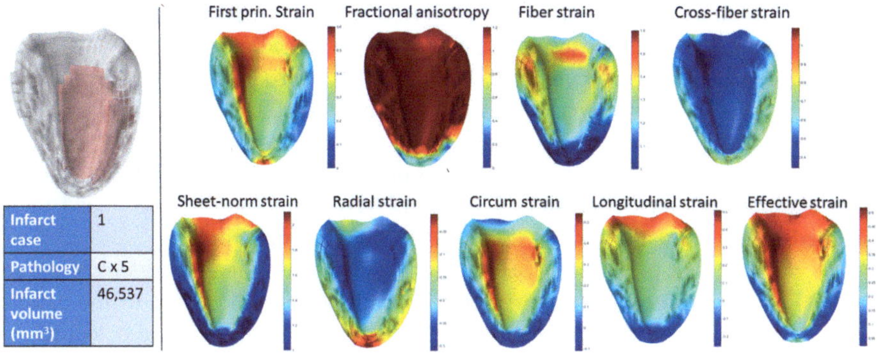

Fig. 6 Visualizations of all strain-based parameters for infarct case number 1

Table 3 Discrimination index values, *S*, for each strain-based parameter

Params	Principal strain	Fractional anisotropy	Fiber stretch	Cross-fiber stretch	Sheet-normal stretch	Radial strain	Circum strain	Longitudinal strain	Effective strain
S	0.08	0.02	**0.15**	0.10	**0.12**	0.05	0.07	0.05	0.08

range, as the healthy left ventricular end-diastolic pressure ranges from 6 to 12 mmHg (Homoud 2008).

Evaluation of Strain-Based Parameters

The resulting geometrical properties of the infarcts are listed in Table 2. Figure 6 shows the end-diastolic color map for each strain-based parameter for one pathological case. The resulting discrimination index *S* for each strain-based parameter is listed in Table 3.

Discussion

This study details how finite-element models can be used to evaluate the sensitivity of different strain-based parameters in detecting myocardial infarct in LVs. As can be seen in Fig. 6, most of the strain-based parameters were able to visually highlight the envelope of the infarct regions. Furthermore, although the fiber stretch showed the best result in discriminating the synthetic infarct in the diastolic phase, the other strain-based parameters are also somewhat sensitive to the infarct regions.

The results of this study are preliminary and the introduced approach could be improved. The fact that the exact ground truth data are available should be exploited by directly conducting an infarct segmentation method in order to detect the most sensitive strain-based parameters. To serve this purpose, we are developing unsupervised statistical-based learning algorithms (Medrano-Gracia 2015) based on a large database of simulated LV models. This work is currently in progress. Second, systolic simulations should also be investigated. Thus, the most sensitive strain-based parameter(s) can be found for both phases and might result in more accurate infarct delineation. Finally, in vivo pathological cases need to be included in the validation of the infarct identification method.

To conclude, our preliminary results prove that it is promising to identify infarct regions via strain-based parameters derived from the diastolic phase of the cardiac cycle. Although the fiber stretch was found to be the most sensitive in detecting the infarct region, further analysis in this regard is required to reach a more conclusive result.

Acknowledgements G.K. Rumindo is supported by the European Commission Horizon 2020 Marie Sklodowska-Curie European Training Network VPH-CaSE (www.vph-case.eu), Grant Agreement No 642612. This work was performed within the LABEX PRIMES (ANR-11-LABX-0063) of Université de Lyon, within the program "Investissements d'Avenir" (ANR-11-IDEX-0007) operated by the French National Research Agency (ANR).

References

ABAQUS (2011) ABAQUS Documentation, Providence, RI, USA: Dassault Systemes

Soleimanifard S et al (2010) Identification of myocardial infarction using three-dimensional strain tensor fractional anisotropy. In: Proceedings of the IEEE international symposium on biomedical imaging, pp 468–471

Medrano-Gracia P et al (2015) Statistical shape modeling of the left ventricle: myocardial infarct classification challenge. In: MICCAI-STACOM

Allman KC, Shaw LJ, Hachamovitch R, Udelson JE (2002) Myocardial viability testing and impact of revascularization on prognosis in patients with coronary artery disease and left ventricular dysfunction: a meta-analysis. J Am Coll Cardiol 39(7):1151–1158

Doltra A et al (2013) Emerging concepts for myocardial late gadolinium. Curr Cardiol Rev 9 (3):185–190

Duchateau N et al (2016) Infarct localization from myocardial deformation: prediction and uncertainty quantification by regression from a low-dimensional space. IEEE Trans Med Imaging 35(10):2340–2352

Elahi MM et al (2014) One problem two issues! Left ventricular systolic and diastolic dysfunction in aortic stenosis. Ann Transl Med 2(1):10

Genet M et al (2014) Distribution of normal human left ventricular myofiber stress at end diastole and end systole: a target for in silico design for heart failure treatments. J Appl Physiol 117:142–152

Genet M et al (2015) A novel method for quantifying smooth regional variations in myocardial contractility within an infarcted human left ventricle based on delay-enhanced magnetic resonance imaging. J Biomech Eng 137(8):081009

Goette MJW et al (2001) Quantification of regional contractile function after infarction: strain analysis superior to wall thickening analysis in discriminating infarct from remote myocardium. J Am Coll Cardiol 37(3):808–817

Holzapfel GA, Ogden RW (2009) Constitutive modelling for passive myocardium: a structurally based framework for material characterization. Philos Trans R Soc A 367:3445–3475
Homoud MK (2008) Introduction to vardiovascular pathophysiology. http://ocw.tufts.edu/data/50/636804.pdf. Accessed 6 Dec 2016
Klotz S et al (2006) Single-beat estimation of end-diastolic pressure–volume relationship: a novel method with potential for noninvasive application. Am J Physiol Heart Circ Physiol 291(1): H403–H412
Lee WN et al (2008) Preliminary validation of angle-independent myocardial elastography using MR tagging in a clinical setting. US Med Biol 34(12):1980–1997
Popovic A et al (1996) Changes in passive mechanical stiffness of myocardial tissue with aneurysm formation. Am J Cardiol 77(5):361–364
Prince JL, McVeigh ER (1992) Motion estimation from tagged MR image sequences. IEEE Trans Med Imaging 11(2):238–249
Schuster A et al (2015) Cardiovascular magnetic resonance feature-tracking assessment of myocardial mechanics: intervendor agreement and considerations regarding reproducibility. Clin Radiol 70(9):989–998
Streeter DD et al (1969) Fiber orientation in the canine left ventricle during diastole and systole. Circ Res 24:339–347
Warren SE, Grossman W (1991) Prognosis in heart failure: is systolic or diastolic dysfunction more important? Herz 1:324–329
Young JM et al (2010) Automatic generation of user material subroutines for biomechanical growth analysis. J Biomech Eng 132(10):104505
Bhat A. et al (2016) Myocardial viability: from proof of concept to clinical practice. Cardiol Res Pract 1020818

Adipogenesis of 3T3L1 Cells Subjected to Tensile Deformations Under Various Glucose Concentrations

Maayan Lustig, Lisa Mor-Yossef Moldovan, Amit Gefen and Dafna Benayahu

Abstract Glucose transport in fat cells results in accumulation of triglycerides in lipid droplets and is regulated by insulin. When a fat tissue becomes insulin-resistant, glucose transport into the cells is impaired and results in Type 2 diabetes. The lipid droplets accumulation is part of the adipogenesis differentiation and metabolism. In the current study, we monitored the adipogenesis of 3T3-L1 cultured cells in high and low glucose concentrations, while the cells were exposed to different substrate rigidity and tensile deformation. Phase contrast images were taken along the adipogenesis process and were analyzed by a new MATLAB image processing algorithm, based on a previous code written in our group (Levy in Annals of biomedical engineering 40:1052–1060, 2012). The new algorithm follows cell differentiation (cell size and morphology and nucleus size) and lipid accumulation (number of lipid droplets per cell and their radius). Complementary, we analyzed by immunofluorescence (IF) the molecular expression of PPARγ, a transcription factor, along with DNA staining by DAPI and Lamin A/C for the nucleus organization. The results indicate that high glucose concentration and substrate tensile strains delivered to adipocytes accelerate their lipid production. In addition, the cell and nucleus area and cell morphology change during the differentiation process.

Introduction

Mesenchymal cells differentiate into various lineages including adipocytes. The adipocyte fate that is visualized by morphological changes of cells, is also associated with lipid droplets (LD) accumulation as part of the adipogenesis

M. Lustig (✉) · A. Gefen
Department of Biomedical Engineering, Faculty of Engineering,
Tel Aviv University, Tel Aviv, Israel
e-mail: maay7@mail.tau.ac.il

L. Mor-Yossef Moldovan · D. Benayahu
Department of Cell and Developmental Biology, Sackler School of Medicine,
Tel Aviv University, Tel Aviv, Israel
e-mail: dafnab@post.tau.ac.il

A. Gefen and D. Weihs (eds.), *Computer Methods in Biomechanics and Biomedical Engineering*, Lecture Notes in Bioengineering,
DOI 10.1007/978-3-319-59764-5_19

metabolism. The LD consists of triglycerides that are formed from internalization of glucose in a process regulated by insulin.

When adipocytes become insulin-resistant, glucose transport into the cells is impaired and results in Type-2 diabetes which is characterized by recurrent or persistent high blood sugar concentrations. In normal condition, the fasting blood glucose concentration should be below 100 mg/dL and diabetes mellitus condition is defined by fasting blood glucose concentration above 126 mg/dL (World Health Organization 2006). In this study, we monitored the adipogenesis of 3T3-L1 cultued cells in high and low glucose concentrations, while the cells were exposed to different ubstrate tensile deformations, based on prior work which revealed that adipogenesis is accelerated in cultures subjected to static, large substrate tensile strains (Shoham et al. 2012; Levy et al. 2012).

Materials and Methods

The differentiation of 3T3-L1 cells was monitored for three weeks, while the cultures were assigned to two experimental conditions

i. Low (100 mg/dL) and high glucose (450 mg/dL) concentrations in the growth medium, to simulate normal and diabetes mellitus conditions, respectively (World Health Organization 2006) (Fig. 1a).

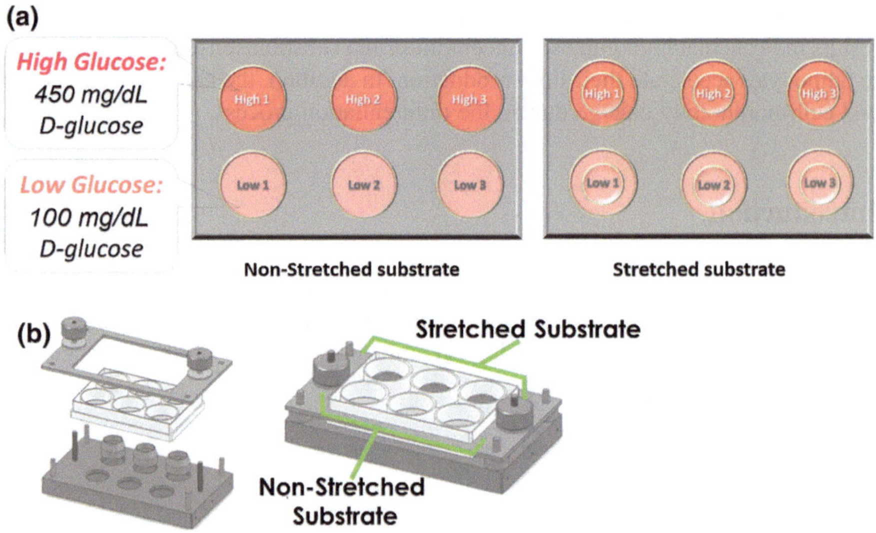

Fig. 1 **a** The experimental setup consisted of two 6-well plates with elastic substrate, one plate was subjected to tensile strain. Each plate contained triplicates of high glucose concentration and low glucose concentration cultures. **b** The cell stretching device, disassembled (*right*) and assembled (*left*)

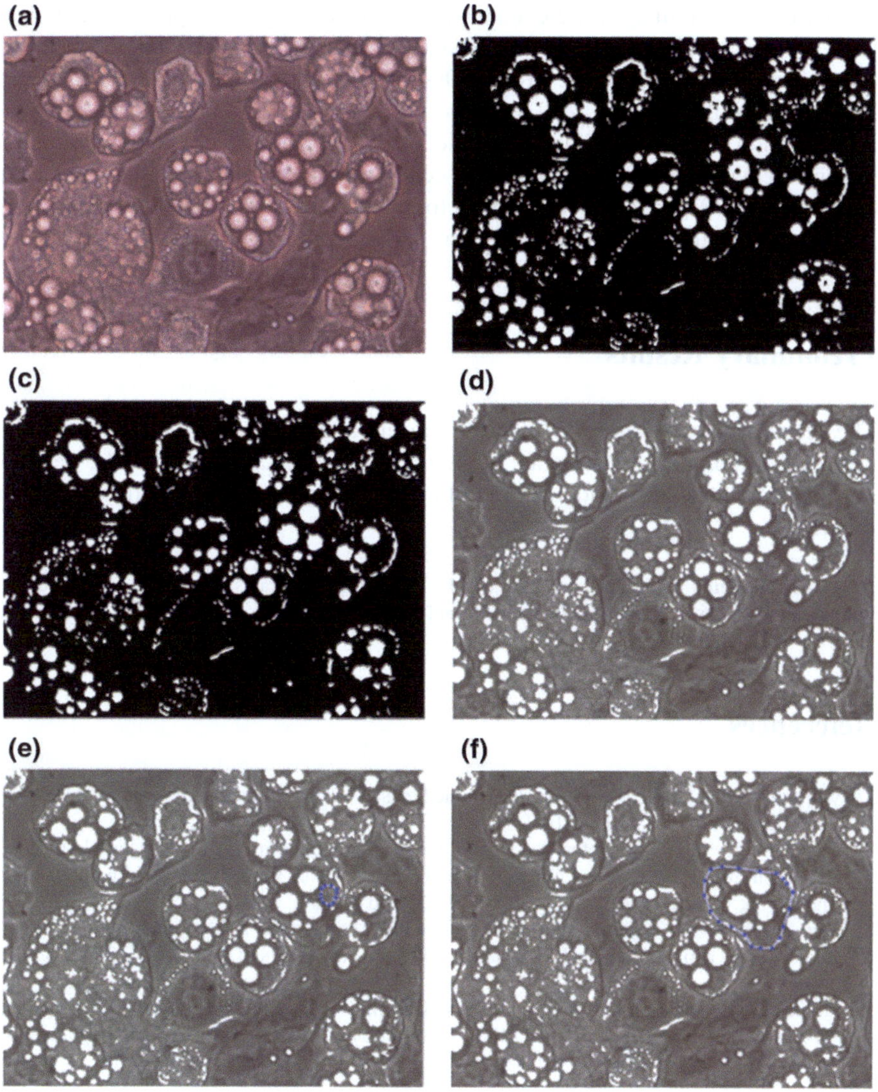

Fig. 2 Main steps in the image proccesing algorithm used for monitoring adipocytes' differentiation. **a** The original image. **b** The image is filtered for random noise and converted to a binary image. **c** Artifact cavities in the lipid droplets are filled. **d** Overlay of the binary image and the grayscale processed image. **e** An individual nucleus is selected for further calculations. **f** An individual cell is selected for further calculations

ii. Cultures were subjected to different tensile strains by employing static stretching on elastic substrate, compared to cultures seeded on a non-stretched elastic substrate (Fig. 1b).

Phase contrast images were taken along the adipogenesis process and were analyzed by an advance MATLAB image processing algorithm, based on a previous code developed in our group (Or-Tzadikario et al. 2010). This algorithm follows cell differentiation (cell size and morphology and nucleus size) and lipid accumulation (number of lipid droplets per cell and their radius) (Fig. 2). Complementary, using immunofluorescence (IF) we analyzed the molecular expression of PPARγ, a transcription factor, along with DNA staining by DAPI and Lamin A/C for the nucleus organization.

Preliminary Results

The results indicate that high glucose concentration and substrate tensile strains delivered to adipocytes accelerate their lipid production. In addition, the cell and nucleus projected area and cell morphology changes during the differentiation process.

Acknowledgements This research work was supported by grants from the Israel Science Foundation (Nos. 611/12 and 1266/16, A.G. and D.B.).

References

Levy A et al (2012) Large, but not small sustained tensile strains stimulate adipogenesis in culture. Ann Biomed Eng 40(5):1052–1060

Or-Tzadikario S, Sopher R, Gefen A (2010) Quantitative monitoring of lipid accumulation over time in cultured adipocytes as function of culture conditions: toward controlled adipose tissue engineering. Tissue Eng Part C: Methods 16(5):1167–1181

Shoham N et al (2012) Static mechanical stretching accelerates lipid production in 3T3-L1 adipocytes by activating the MEK signaling pathway. Am J Physiol Cell Physiol 302(2):C429–C441

World Health Organization (2006) Definition and diagnosis of diabetes mellitus and intermediate hyperglycemia: report of a WHO/IDF consultation (PDF). World Health Organization, Geneva, p 21. ISBN 978-92-4-159493-6

Computer Simulation of Electroporation and Drug Transport Through Membranes

Nenad Filipovic, Igor Saveljic and Irena Tanaskovic

Abstract In this study, experimental electroporation model with human aorta tissue is compared with computational modeling. The segments in native state of the aorta are treated by electroporation method through a series of electrical impulses from 50 to 2500 V/cm. The Pennes Bioheat equation is used to solve heat transfer problems. Different conductivity values are used in order to fit the experimental results. It has been shown that there are a smaller number of vascular smooth muscle cells (VSMC) nuclei at the tunica media, while the elastic fibers morphology is maintained 24 h after electroporation. Additionally we studied with computational model of plaque formation and progression the reduction of the plaque size with electroporation. The initial results have been shown plaque reduction for carotid artery case. Future studies are necessary for design of a new device for in vivo ablation with electroporation of plaque stenosis.

Introduction

Electroporation includes application of electrical pulses with charging of lipid bilayer membranes and rapid localized structural rearrangements within the membrane. Irreversible electroporation (IRE) is electrical field effect applied across a cell which destabilizes electric potential across the cell outer membrane and causes formation of permanent defects in the lipid bilayer. It leads to the changes in cell homeostasis and cell death (Deng and Liu 2001; Sale and Hamilton 1967; Hamilton and Sale 1967). Due to a very short duration, IRE could ablate substantial volumes of tissue without thermal effects (Sale and Hamilton 1968).

There is an increase in the conductivity throughout the electroporation pulse. Permeabilization of the cell membrane caused with electroporation increases the

N. Filipovic (✉) · I. Saveljic
Faculty of Engineering, University of Kragujevac, 34000 Kragujevac, Serbia
e-mail: fica@kg.ac.rs

I. Tanaskovic
Faculty of Medical Sciences, University of Kragujevac, 34000 Kragujevac, Serbia

A. Gefen and D. Weihs (eds.), *Computer Methods in Biomechanics and Biomedical Engineering*, Lecture Notes in Bioengineering,
DOI 10.1007/978-3-319-59764-5_20

conductivity of the tissue. Maor et al. (2007) found that vascular smooth muscle cells (VSMC) nuclei begins to reduce which potentially can lead to possible plaque reduction. For plaque reduction model we described mass transport of LDL through the wall and the simplified inflammatory process. The Navier–Stokes equations govern the blood motion in the lumen, the Darcy law is used for model blood filtration, Kedem–Katchalsky equations (Kedem and Katchalsky 1961, 1958) for the solute and flux exchanges between the lumen and the intima. The system of three additional reaction–diffusion equations that models the inflammatory process and lesion growth model in the intima is used (Filipovic et al. 2012, 2013).

We investigated the ablation effect on the human aorta tissue during electroporation and use this computational method for virtual application for plaque reduction. The experimental procedure for electroporation is described (Filipovic et al. 2016). Computational model for electroporation and plaque reduction is described. The obtained results could support idea that electroporation could be a potential tool for plaque sites reduction.

Materials and Methods

Histology Procedure for Electroporation

We used samples during the autopsy of 35 male cadavers, 40–65 years old, who died of nonvascular samples for histology analysis. The degree of atherosclerosis of the samples was determined according to the classification of the American Heart Association Committee on Vascular Lesions of the Council of Atherosclerosis. All analyzed samples were in preatheroma stage (stage 3).

The segments in native state are treated by electroporation method with the device: ECM 399 electroporation system (BTX Harvard apparatus, Massachusetts, the United States), through a series of electrical impulses of low voltage (LV) of 50 to 500 LV and high voltage (HV) from 100 to 2500 HV (Filipovic et al. 2016).

For each analyzed sample, one control of the same age was used, without the leakage of electrical impulses.

Computational Model of Electroporation

The Pennes Bioheat equation is used to solve heat transfer problems. This equation has the following form:

$$\nabla \cdot (k\nabla T) + p = \rho c_p \frac{\partial T}{\partial t}, \tag{1}$$

where k is the thermal conductivity of the tissue, T is the temperature, ρ is the tissue density, and c_p is the heat capacity of the tissue.

Due to electroporation and temperature, the conductivity changes were used to calculate the dynamic conductivity according to the following equation:

$$\sigma = \sigma_0[1 + f(E - E_0, E_1) + \alpha(T - T_0)], \tag{2}$$

where σ_0 is the baseline conductivity, f is a smoothed heavyside function with a continuous second derivative that ensures the numerical solution convergence. This function changes from zero to one when $E - E_0 = 0$ over the range E_1. The baseline tissue conductivity was set to 0.067 S/m.

Modeling of Plaque Formation and Progression

The blood flow in lumen domain is simulated by the three-dimensional Navier–Stokes equations, together with the continuity equation. Mass transfer in the blood lumen is coupled with the blood flow and modeled by the convection–diffusion equation. Mass transfer in the arterial wall is coupled with the transmural flow and modeled by the convection–diffusion–reaction equation. LDL transport in lumen of the vessel is coupled with Kedem–Katchalsky equations (Hamilton and Sale 1967; Sale and Hamilton 1968).

We are using the incremental-iterative procedure to treat the convective diffusion terms for LDL transport.

The inflammatory process was solved using three additional reaction–diffusion partial differential equations (Maor et al. 2007; Kedem and Katchalsky 1961):

$$\begin{aligned} &\partial_t Ox = d_1 \Delta Ox - k_1 Ox \cdot M \\ &\partial_t M + div(v_w M) = d_2 \Delta M - k_1 Ox \cdot M + S/(1+S) \\ &\partial_t S = d_3 \Delta S - \lambda S + k_1 Ox \cdot M + \gamma(Ox - Ox^{\mathrm{thr}}), \end{aligned} \tag{3}$$

where Ox is the oxidized LDL or c_w—the solute concentration in the wall; M and S are concentrations in the intima of macrophages and cytokines, respectively; d_1, d_2, d_3 are the corresponding diffusion coefficients; λ and γ are degradation and LDL oxidized detection coefficients; and v_w is the inflammatory velocity of plaque growth, which satisfies Darcy's law and continuity equation (Maor et al. 2007; Kedem and Katchalsky 1961):

$$v_w - \nabla \cdot (p_w) = 0 \tag{4}$$

$$\nabla v_w = 0 \tag{5}$$

in the wall domain. Here, p_w is the pressure in the arterial wall.

Results

Voltage distribution for prescribed voltage on the electrode 2500 V is shown in the Fig. 1.

Electrical field distribution for applied electroporation 2500 V/cm is presented in Fig. 2. We used the values of the tissue heat capacity ($c_p = 3.6$ kJ kg^{-1} K^{-1}), thermal conductivity ($k = 0.512$ W m^{-1} K^{-1}) and density ($\rho = 1050$ kg m^{-3}).

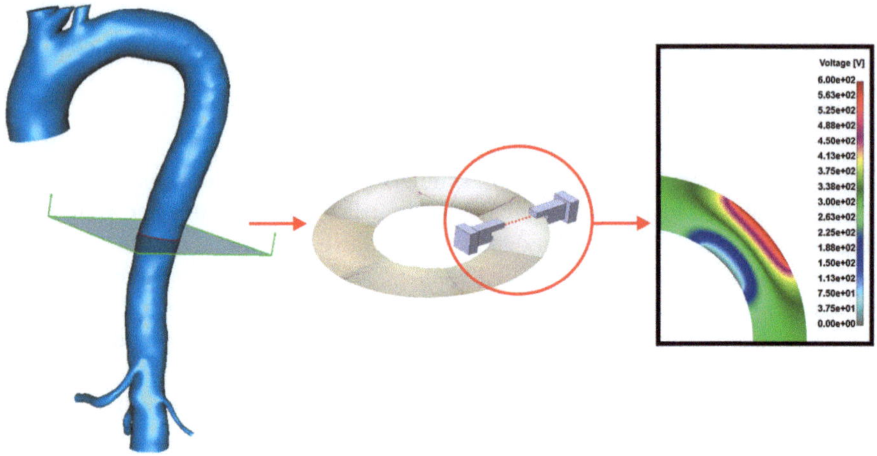

Fig. 1 Voltage distribution for applied electrode—2500 V/cm

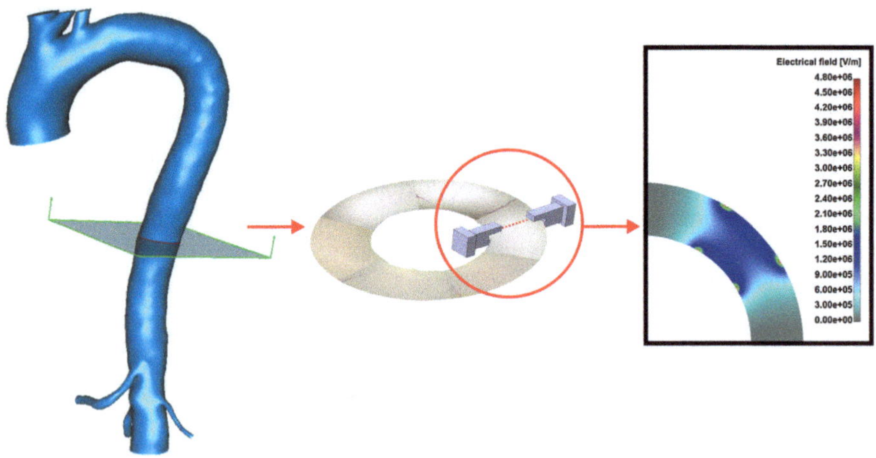

Fig. 2 Electrical field distribution for applied electroporation—2500 V/cm

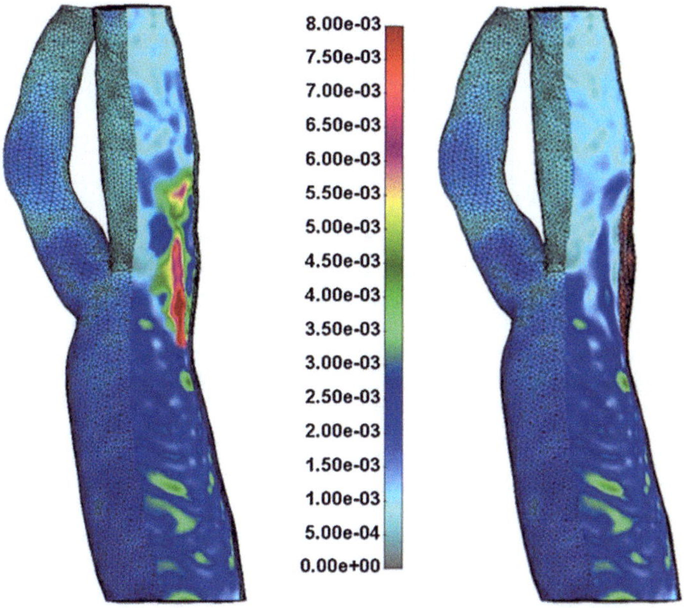

Fig. 3 Reducing plaque concentration with virtual electroporation for carotid artery. The units are (mg/ml)

We simulated virtual reducing of the plaque concentration with electroporation on the specific carotid artery case. The result for plaque concentration before and after electroporation process has been shown in the Fig. 3.

Discussion and Conclusions

We treated with electroporation the segments of the human aorta in native state. The electrical impulses were low voltage (LV) of 50–500 LV and high voltage (HV) from 100 to 2500 HV. During experimental procedure it has been shown that there is smaller number of vascular smooth muscle cells (VSMC) nuclei at the tunica media while the elastic fibers morphology is maintained.

The computational model for electroporation includes transfer problem coupled with electrical field described by Pennes Bioheat equation. The conductivity changes due to electroporation and temperature were used to calculate the dynamic conductivity.

The reduction of VSMC could be potential tool plaque decrease and therefore the reduction of the stenosis. Influence of electroporation on the carotid artery example case was virtually investigated where the size of the plaque was reducing.

Acknowledgements This study was funded by grants from Serbian Ministry of Education, Science and Technological Development III41007, ON174028 and HORIZON2020 689068 SMARTool project.

References

Deng ZS, Liu J (2001) Blood perfusion-based model for characterizing the temperature fluctuations in living tissue. Phys A Stat Mech Appl 300:521–530

Filipovic N, Rosic M, Tanaskovic I, Milosevic Z, Nikolic D, Zdravkovic N, Peulic A, Fotiadis D, Parodi O (2012) ARTreat project: three-dimensional numerical simulation of plaque formation and development in the arteries. IEEE Trans Inf Technol Biomed 16(2):272–278

Filipovic N, Teng Z, Radovic M, Saveljic I, Fotiadis D, Parodi O (2013) Computer simulation of three dimensional plaque formation and progression in the carotid artery. Med Biol Eng Comput. doi:10.1007/s11517-012-1031-4

Filipovic N, Saveljic I, Jovicic N, Tanaskovic I, Zdravkovic N (2016) Computational and experimental model of electroporation for human aorta. Acta Bioeng Biomech. doi:10.5277/ABB-00444-2015-02

Hamilton WA, Sale JH (1967) Effects of high electri fields of microorganisms. II. Mechansims of action of the lethal effect. Biochem Biophys Acta 148:789–800

Kedem O, Katchalsky A (1958) Thermodynamic analysis of the permeability of biological membranes to non-electrolytes. Biochim Biophys 27:229–246

Kedem O, Katchalsky A (1961) A physical interpretation of the phenomenological coefficients of membrane permeability. J Gen Physiol 45:143–179

Maor E, Ivorra A, Leor J, Rubinsky B (2007) The effect of irreversible electroporation on blood vessels. Technol Cancer Res Treat 6(4):307–312. ISSN 1533-0346

Sale JH, Hamilton WA (1967) Effect of high electrical fields on microorganisms. I. Killing of bacteria and yeast. Biochim Biophys Acta 148:781–788

Sale AJH, Hamilton WA (1968) Effects of high electri fields on microorganisms III. Lysis of erythrocytes and protoplasts. Biochim Biophys Acta 163:37–43

Development of a Computational Model to Aid Prediction of Neurosurgical Brain Shift

N.J. Bennion, M. Potts, A.D. Marshall, S. Anderson and S.L. Evans

Abstract Stereotactic procedures are an increasingly common tool for the diagnosis and treatment of neurological disorders. Common surgeries reliant on a stereotactic reference frame include Deep Brain Stimulation, Stereoelectroencephalography, Stereobiopsy, and high precision intraparenchymal drug delivery. Introduction: Stereotactic neurosurgical procedures are planned and carried out per preoperative medical images in a fixed reference frame. Loss of cerebrospinal fluid and a variety of other factors lead to a displacement of the anatomical target from the stereotactic coordinates, known as brain shift. Aims: To develop a computational model to aid in the understanding and prediction of gravity induced brain shift based on patient repositioning. Methods: The MNI ICBM152 Average Brain Stereotaxic Registration Model was manually segmented and meshed in the Simpleware Scan IP software package. Using FEBio, suitable constitutive models were applied to each region. The model was then loaded to simulate supine-to-prone repositioning. Results: Displacement reached a maximum of approximately 2.4mm, with cortical displacement being concentrated in anterior regions. Conclusions: With good initial results, the future applications of this method appear promising.

This research is funded by the EPSRC and Renishaw Plc. as part of an iCase Studentship. The authors acknowledge the contributions of Rob Harrison, Renishaw Plc.

N.J. Bennion (✉) · M. Potts · A.D. Marshall · S.L. Evans
School of Engineering, Cardiff University, Cardiff CF24 3AA, UK
e-mail: bennionn@cardiff.ac.uk

S. Anderson
School of Engineering, Renishaw Plc, New Mills, Wotton-under-Edge, Gloucestershire GL12 8JR, UK

A. Gefen and D. Weihs (eds.), *Computer Methods in Biomechanics and Biomedical Engineering*, Lecture Notes in Bioengineering,
DOI 10.1007/978-3-319-59764-5_21

Introduction

Stereotactic procedures are an increasingly common tool for the diagnosis and treatment of neurological disorders. Common surgeries reliant on a stereotactic reference frame include Deep Brain Stimulation (DBS), Stereoelectroencephalography, Stereobiopsy and high precision intra-parenchymal drug delivery (Elias et al. 2007; Lewis et al. 2016; Jackson et al. 2001). DBS is used to treat the symptoms of movement disorders such as Parkinson's disease, dystonia, and essential tremors (Hamze et al. 2015). As its name suggests, an electrode is inserted through the skull into a specific region of the deep brain. The surgeon will decide the best regions to aim for based on the condition that requires treatment and any patient specific variation. The subthalamic nucleus (STN), globus pallidus internus (GPi), ventral intermediate thalamus (VIM), and pedunculopontine nucleus (PPN) are common in DBS procedures (Kalia et al. 2013). Given the very small size of these structures, the efficacy of this treatment depends on sub-millimetre accuracy of electrode placement.

Surgeons use pre-operative patient images to identify the electrode trajectory and insertion depth required to reach anatomical targets. It has, however, been long understood that the process of inserting an electrode into the brain leads to a non-rigid deformation known as brain shift (Elias et al. 2007). The existence of brain shift in procedures requiring such accuracy has led to the development of various techniques such as microelectrode recording and intraoperative MRI to account for this phenomenon and to ensure adequate placement of electrodes. Although effective, these additional compensatory measures add time and complexity to the operation (Amirnovin et al. 2006; Ivan et al. 2014).

It has been clearly demonstrated that losing large amounts of cerebrospinal fluid (CSF) can lead to large deformations of both the cortical surface and deep brain (Elias et al. 2007). However, even when surgical protocols developed specifically to limit CSF loss are in place, deep brain displacement of a surgically significant scale is still observed. It is therefore reasonable to conclude that brain shift is not simply a function of CSF loss. As such, a wide range of patient specific factors must be important. Although studies have attempted to assess the impact of some measurable forms of intra-patient variations, such as cerebral atrophy (Azmi et al. 2011), a definitive understanding has not yet been achieved.

In order to develop a method of accounting for brain shift on a patient specific basis, a better understanding of brain shift which occurs with no surgical intervention is required. This is known as anatomical brain shift and occurs in everyone simply with repositioning of the head with respect to gravity. Anatomical brain shift has been studied by MR imaging of patients lying in prone, supine and lateral positions (Hill et al. 1998; Schnaudigel et al. 2010; Monea et al. 2012). Results from such studies have again highlighted that clinically significant displacement of both the deep brain and cortical surface can be expected even with no surgical intervention.

The purpose of this study is to develop a detailed computational model to predict anatomical brain shift. This model can then be used to further our understanding of patient specific geometric and tissue property variation, and its impact on brain shift.

Materials and Methods

Image Segmentation and Model Geometry

Image segmentation was performed manually using Simpleware ScanIP (Synopsys, Mountain View, USA). In order to avoid patient specific anatomical variation the MNI ICBM152 Average Brain Stereotaxic Registration Model (McConnell Brain Imaging Centre, Montreal Neurological Institute, McGill University) was used.

Throughout the early stages of development, models were segmented with increasing levels of anatomic detail. In particular, simplified models suggested that the pia-arachnoid complex (PAC) and dural septa offered considerable support to the brain. This highlighted the need to balance such detail with that for a computationally stable mesh. As a result, the model was segmented into the following parts: skull, brain, ventricular system, pia-arachnoid complex, dural septa, and major sinuses.

The skull was included only to act as an outer limit and was assumed to be rigid in the final model. To improve computational stability some smoothing and manual inclusion/removal of certain areas was also required. Offering important structural support, the tentorium cerebelli and falx cerebri were segmented together with the sinuses adjacent to them. Whether it is correct to assume these structures do not deform significantly will be tested in the future. In this model, the dura mater was assumed to be rigidly attached to the skull and therefore did not play a structural role. The PAC, which surrounds the brain, was segmented as one layer, but split into left, right and inferior sections.

Material Parameters

The segmented geometry was imported into FEBio (Maas et al. 2012) for finite element analysis. Table 1 outlines the FEBio material parameters that were chosen for each region.

Table 1 Material parameters used in the final computational model

Structure	Material model	Material parameters	Elements
Brain	Mooney–Rivlin	$c1 = 0.28$, $c2 = 333$ (Mihai et al. 2015), $k = 1{,}000{,}000$ Pa, $\rho = 1040$ kg/m^3 (Cala et al. 1981)	766,611
PAC	Transversely isotropic Mooney–Rivlin	$c1 = 10$, $c2 = 10$, $c3 = 1.8$, $c4 = 175$, $c5 = 80{,}000$ Pa, lam_max = 1.01 $k = 10{,}000$ Pa (Jin et al. 2006; Jin et al. 2007; Jin et al. 2014; Jin et al. 2011), $\rho = 1007$ kg/m^3 (Levin et al. 1981)	645,023
Ventricles	Isotropic elastic	$E = 1000$ Pa, $v = 0.49$	28,843
Dural septa	Isotropic elastic	$E = 30{,}000{,}000$ Pa (van Noort et al. 1981), $v = 0.49$	158,442

As noted above, the PAC regions were intended to be fiber stiffened in the out of plane vector in order to replicate the tethering action of the arachnoid trabeculae. To achieve this, the PAC structure was split into three regions, with a spherical fiber center assigned to the central point of each hemisphere and the cerebellum. A linear tetrahedral mesh was also generated and refined in Simpleware ScanIP.

Loading

It is not possible to image the brain in an unloaded state as it is always subject to the force of gravity in the initial imaging position. As a result, the model must account for the force difference between the prone and supine positions as opposed to those generated in either position alone. Given that this is a quasi-static analysis with no surgical interventions, the only two loads on the system are gravity and buoyancy force generated by the density difference between brain tissue and CSF. This buoyancy force can be implemented as a pressure on the outside surface of the brain.

The positional change means that the gravitational body load acting is 18.62 m/s^2.

Taking the initial supine imaging position, the pressure P at any point is equal to:

$$P = \rho g h,$$

where ρ is density, g is the gravitational constant and h is the distance parallel to the gravitational field from the top of the fluid body to the point. Taking the anterior limit of the intracranial volume to be h_0 and the posterior limit to be h_1, the pressure distribution at any point in the supine position P_s is:

$$P_s = \rho g(h_0 - h_i),$$

where h_i is the position of the point in the gravitational axis. Similarly, the pressure distribution in the prone position P_p is given by:

$$P_p = \rho g(h_i - h_1)$$

As such, the pressure difference ΔP at any point when moving from supine to prone is

$$\Delta P = P_p - P_s$$

$$\Delta P = \rho g(h_i - h_1) - \rho g(h_0 - h_i)$$

$$\Delta P = -\rho g(h_1 + h_o - 2h_i)$$

This pressure was calculated and applied to every element surface making up the exterior of the brain.

Results

The results of the computational model are shown in Fig. 1. Deformation was concentrated around the corpus callosum and anterior portions of the lateral and third ventricles. In this region the total displacement reached a maximum of 2.36 mm and was mainly a result of component vectors in the gravitational axis.

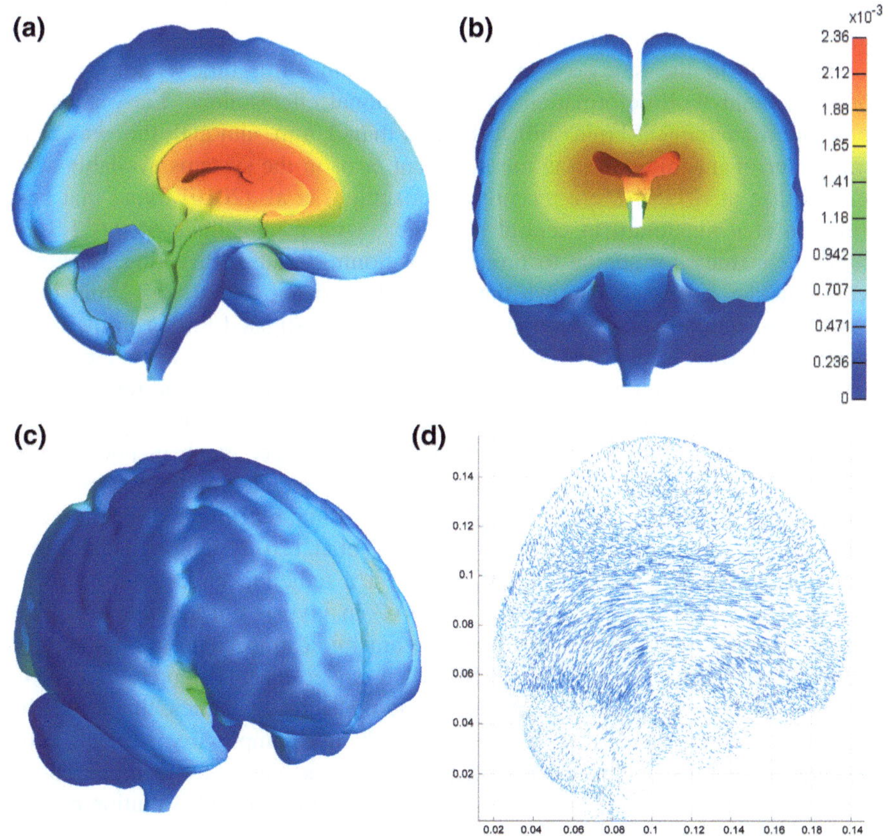

Fig. 1 Scalar deformation in meters that was predicted as a result of supine to prone repositioning (**a**) in a midline section in the sagittal plane, **b** in a coronal section and, **c** on the cortical surface. **d** Is a vector plot visualizing 1/20 of the nodal vectors from the brain and other structures

Cortical displacement was greater in anterior portions of the brain, remaining less than 1 mm in all regions.

Discussion

Although many groups have attempted to model brain deformation for cases where large scale resection of tissue is required, far fewer have investigated predictions for stereotactic cases. Bilger et al. (2011, 2014a, b) also accepted that prediction of CSF loss was impossible. Instead, the authors employed worse case boundary conditions to generate a risk volume that must be avoided to ensure safe insertion of the electrode. This and subsequent related works looked at automation of this process and intra-operative registration of the computational model to a deformed image. These models use a physics-based approach incorporating similar boundary conditions. However, the slight difference in application means comparison of the models directly is not meaningful.

Compared to the study of intraoperative brain shift, little is known about non-rigid brain-skull deformation due to positional effects alone. Early investigations found anatomical brain shift to be less than 1 mm, although this was in the same order of magnitude as the measurement error in the systems available at the time (Hill et al. 1998). More recently, Schnaudigel et al. (2010) found that brain shift was at a maximum in central structures, with a magnitude of 0.6–1.3 mm across their study for supine to prone repositioning. Although taking place relatively quickly, it was also shown that the brain had not settled by 12 min in the scanning position; a consideration which is to be investigated further in the future. In contrast, Monea et al. (2012) found the greatest deformations to be on the cortical surface with a maximum of 7.86 mm for the same prone to supine repositioning. The source of the difference between these two studies is unclear.

The level of deformation found in this study fell very much in line with the results of Schnaudigel et al. (2010). Cortical displacement was limited, but more significant in the anterior half of the brain. The absolute magnitude of displacement was not in the region of those reported by Monea et al. (2012), although distribution was similar. The results of this model suggest a tethering effect of the PAC in tension. The compressive stiffness of the PAC in vivo is not known. It is likely comprised of a slight contribution from the arachnoid trabeculae in buckling, but mainly from redistribution of the CSF when under increased pressure. As the current model does not include any computational fluid dynamics, this aspect has been simplified to standard solid elements with low compressive stiffness and a less-than-incompressible bulk modulus. Finding precise values for these parameters that are required to provide an accurate representation of this redistribution requires further investigation.

Due to a lack of a consensus and large degrees of intra-patient variation seen in the literature, more rigorous validation of this model is not currently possible. Regardless of this, the current study has shown that the combination of detailed geometry and structure specific material parameterization can offer new insights into the mechanics of anatomical brain deformation. With continued development, this approach has the potential to offer significant improvements to surgical planning.

References

Amirnovin R, Williams ZM, Cosgrove GR, Eskandar EN (2006) Experience with microelectrode guided subthalamic nucleus deep brain stimulation. Neurosurgery. 58(1 Suppl):ONS96–ONS102 discussion ONS96

Azmi H, Machado A, Deogaonkar M, Rezai A (2011) Intracranial air correlates with preoperative cerebral atrophy and stereotactic error during bilateral STN DBS surgery for Parkinson's disease. Stereotact Funct Neurosurg 89(4):246–252

Bilger A, Dequidt J, Duriez C, Cotin S (2011) Biomechanical simulation of electrode migration for deep brain stimulation. Med Comput Comput Assist Interv Miccai 2011 6891:339–346

Bilger A, Bardinet E, Fernandez-Vidal S, Duriez C, Jannin P, Cotin S (2014a) Intra-operative registration for stereotactic procedures driven by a combined biomechanical brain and CSF model. Lecture Notes in Computer Science, vol 8789, pp 76–85

Bilger A, Duriez C, Cotin S (2014b) Computation and visualization of risk assessment in deep brain stimulation planning. Stud Health Technol Inf 196:29–35

Cala LA, Thickbroom GW, Black JL, Collins DW, Mastaglia FL (1981) Brain density and cerebrospinal fluid space size: CT of normal volunteers. AJNR Am J Neuroradiol 2(1):41–47

Elias WJ, Fu KM, Frysinger RC (2007) Cortical and subcortical brain shift during stereotactic procedures. J Neurosurg 107(5):983–988

Hamze N, Bilger A, Duriez C, Cotin S, Essert C (2015) Anticipation of brain shift in deep brain stimulation automatic planning. Conf Proc IEEE Eng Med Biol Soc 2015:3635–3638

Hill DLG, Maurer CR, Maciunas RJ, Barwise JA, Fitzpatrick JM, Wang MY (1998) Measurement of intraoperative brain surface deformation under a craniotomy. Neurosurgery 43(3):514–526

Ivan ME, Yarlagadda J, Saxena AP, Martin AJ, Starr PA, Sootsman WK et al (2014) Brain shift during bur hole-based procedures using interventional MRI. J Neurosurg 121(1):149–160

Jackson RJ, Fuller GN, Abi-Said D, Lang FF, Gokaslan ZL, Shi WM et al (2001) Limitations of stereotactic biopsy in the initial management of gliomas. Neuro Oncol 3(3):193–200

Jin X, Lee JB, Leung LY, Zhang L, Yang KH, King AI (2006) Biomechanical response of the bovine pia-arachnoid complex to tensile loading at varying strain-rates. Stapp Car Crash J 50:637–649

Jin X, Ma C, Zhang L, Yang KH, King AI, Dong G et al (2007) Biomechanical response of the bovine pia-arachnoid complex to normal traction loading at varying strain rates. Stapp Car Crash J 51:115–126

Jin X, Yang KH, King AI (2011) Mechanical properties of bovine pia-arachnoid complex in shear. J Biomech 44(3):467–474

Jin X, Mao H, Yang KH, King AI (2014) Constitutive modeling of pia-arachnoid complex. Ann Biomed Eng 42(4):812–821

Kalia SK, Sankar T, Lozano AM (2013) Deep brain stimulation for Parkinson's disease and other movement disorders. Curr Opin Neurol 26(4):374–380

Levin E, Muravchick S, Gold MI (1981) Density of normal human cerebrospinal fluid and tetracaine solutions. Anesth Analg 60(11):814–817

Lewis O, Woolley M, Johnson D, Rosser A, Barua NU, Bienemann AS et al (2016) Chronic, intermittent convection-enhanced delivery devices. J Neurosci Methods 259:47–56

Maas SA, Ellis BJ, Ateshian GA, Weiss JA (2012) FEBio: finite elements for biomechanics. J Biomech Eng 134(1):011005

Mihai LA, Chin L, Janmey PA, Goriely A (2015) A comparison of hyperelastic constitutive models applicable to brain and fat tissues. J R Soc Interface 12(110):0486

Monea AG, Verpoest I, Vander Sloten J, Van der Perre G, Goffin J, Depreitere B (2012) Assessment of relative brain-skull motion in quasistatic circumstances by magnetic resonance imaging. J Neurotrauma 29(13):2305–2317

Schnaudigel S, Preul C, Ugur T, Mentzel HJ, Witte OW, Tittgemeyer M et al (2010) Positional brain deformation visualized with magnetic resonance morphometry. Neurosurgery 66(2):376–384 discussion 84

van Noort R, Black MM, Martin TR, Meanley S (1981) A study of the uniaxial mechanical properties of human dura mater preserved in glycerol. Biomaterials 2(1):41–45

Coronary Pressure Drop and Arterial Distensibility—Two Dependent Parameters

Oren M. Rotman, Uri Zaretsky, Avraham Shitzer and Shmuel Einav

Abstract Blood pressure drop and arterial distensibility are important, allegedly independent, parameters that are indicative of the coronary arteries patency and atherosclerosis severity. In the present study we show that these two parameters are dependent, allowing to obtain both parameters from a single measurement, and to do so a high-resolution differential pressure measurement system is required. The objective of the study was to unveil the relationship between local fluid pressure drops and tube distensibility, through various scenarios of stenosis severity, tube diameter, and flow rate (for coronary hemodynamic conditions). The investigation was performed using validated Fluid–Structure Interaction (FSI) analysis on silicone mock arteries with intermediate size stenosis (0–65%). Highly accurate pressure drop measurements (±0.05 mmHg) were performed with our in-house fluid-filled double-lumen catheter measurement. The results indicated that our accurate pressure drop measurement method facilitated the differentiation among several levels of the mock arteries' stiffness, with distensibilities ranging from 1 to 7%. Local pressure drops were markedly affected by the mock arteries' stiffness and could be best described using a second-order polynomial function. These changes in pressure drop could be detected even when there was no stenosis present, and FFR values remained insensitive at 1.00. The results indicated the clinical potential of a high accuracy pressure drop-based parameter to be superior to FFR. Such a parameter would provide cardiovascular interventionalists with lesion-specific bio-mechanical and functional data, for improved real-time decision-making in the cath-lab.

O.M. Rotman (✉) · S. Einav
Department of Biomedical Engineering, Stony Brook University, NY, USA
e-mail: orenrotman1@gmail.com

O.M. Rotman · U. Zaretsky · S. Einav
Department of Biomedical Engineering, Tel-Aviv University, Tel-Aviv, Israel

A. Shitzer
School of Mechanical Engineering, Technion IIT, Haifa, Israel

A. Gefen and D. Weihs (eds.), *Computer Methods in Biomechanics and Biomedical Engineering*, Lecture Notes in Bioengineering,
DOI 10.1007/978-3-319-59764-5_22

Introduction

Blood pressure drop and arterial distensibility are important, allegedly independent, parameters that are indicative of the coronary arteries patency and atherosclerosis severity. Fractional Flow Reserve (FFR) is the gold standard coronary pressure drop-based parameter, utilizing coronary trans-stenotic pressure measurement as a surrogate for flow under maximal hyperemia, and is mostly useful for intermediate size stenoses, where coronary angiography sensitivity is limited (Tobis et al. 2007). Impaired distensibility was correlated with atherosclerotic arteries in comparison to healthy arteries (Mokhtari-Dizaji et al. 2006; van Popele et al. 2001), even in sites accompanying occult lesions, which could not be detected by conventional angiography (Nakatani et al. 1995). Over the years, arterial pressure drop and distensibility were studied separately, as two independent parameters. This study hypothesized that arterial pressure drop was dependent on the arterial distensibility. We were seeking to investigate this relationship through a combination of coronary flow rates and diameters, for zero to intermediate size stenoses. The motivation was to develop a capability to obtain both functional and biomechanical information from a single pressure measurement during cardiac catheterization procedure.

Materials and Methods

Simplified coronary arteries were simulated as straight compliant tubes with a concentric 5 French catheter to match our experimental silicone models setup. The coronary models were scaled up to a baseline diameter of 5 mm to keep the simulations identical to the experimental model, utilizing flow and pressure recordings as boundary conditions (B.C.), and subsequently validating the computational results (Rotman et al. 2016). High-resolution pressure drop was measured experimentally as was described in Rotman et al. (2014, 2015). Simulations were first run on zero-stenosis models as the basis for pressure drop–distensibility relationship. This set of simulations included the effect of variability of the tube nominal diameter. Then, four small to intermediate cases of stenosis severity were simulated: 14, 38, 55, and 64% (by cross-sectional area). For every combination of flow and diameter, or flow and stenosis severity, 4–8 cases of distensibility were simulated in the physiological range of 0–8%. Moreover, additional simulations were performed with the intermediate stenoses models to investigate the effect of hyperemic flow variability on the pressure drop, and for comparison with FFR. Post-processing of the pressure drop signals included extraction of the Mean and Maximum values, which were then correlated with the model distensibility, diameter, flow rate, axial location along the model, and stenosis severity. Four marginal cases from the aforementioned simulations were validated experimentally using our catheterization simulator (Rotman et al. 2016).

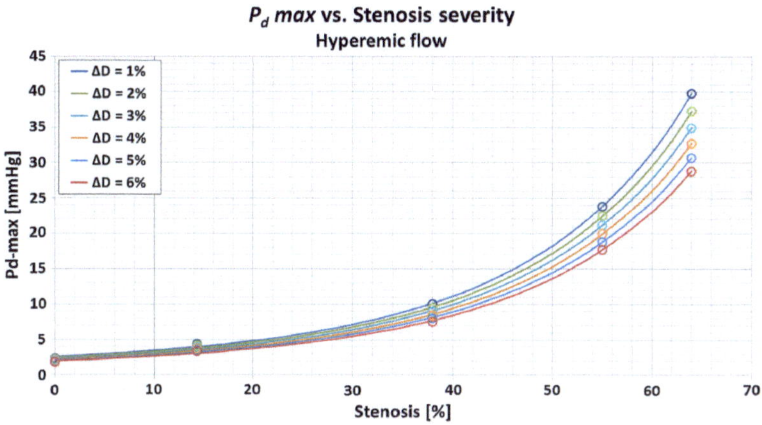

Fig. 1 Plots of P_d-max as a function of the stenosis severity (0–64%) under hyperemic flow, for varying ΔD cases (1–6%), as was calculated in the center of the stenosis. A second-order exponential function was found to best fit these plots

Results

Overall 54 simulation combinations of flow rate, distensibility, and nominal diameter were performed in zero stenosis models. Detailed diagrams were created to describe the correlation of the pressure drop with the aforementioned parameters, and with the axial location along the tube. The later being the result of the pressure waveform decay along the compliant tube model. Thirty-four additional simulations were performed in small to intermediate size stenoses, where pressure drop was measured at seven locations along the stenosis. Analysis was then performed on measurements from the center of the stenosis. Example of the relationship between the peak pressure drop (P_d-max) and the stenosis severity (0–64%) during maximum hyperemia can be seen in Fig. 1, for varying cases of stenosis compliance. These plots were found to be best fitted by using a second-order exponential function $y = e\text{\^{}}(ax\text{\^{}}2 + bx + c)$ ($R^2 > 0.99$). The average sensitivity of P_d-max to a 1% change of ΔD under normal flow conditions was 0.04, 0.10, 0.27, and 0.60 mmHg for stenoses severity of 14, 38, 55, and 64%, respectively. For hyperemic flow the average sensitivity was 0.20, 0.49, 1.23, and 2.19 mmHg for stenoses severities of 14, 38, 55, and 64%, respectively. Additional 12 cases were simulated with the intermediate size stenosis to investigate the effect of variability in hyperemic flow on the measured pressure drop and for comparison with FFR. Plots of $P_d - \max(\text{flow})$ are presented in Fig. 2. This relationship can be approximated as linear, with $R^2 > 0.99$. The results indicate that the pressure drop was positively dependent on the flow rate, and very sensitive to changes thereof. Interestingly, the results show that for a given stenosis case increase of the hyperemic flow rate reduces the FFR, and vice versa. This might lead to a false positive or false negative FFR indication, respectively.

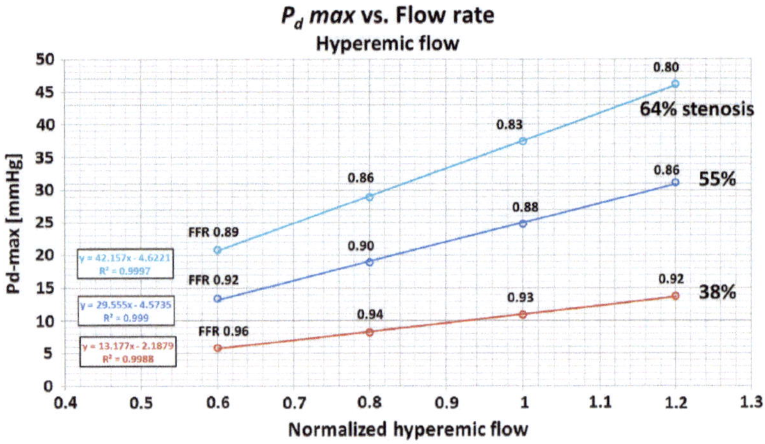

Fig. 2 Plots of P_d-max as a function of the normalized hyperemic flow, at the center of the stenosis, for intermediate size stenosis cases, with linear fitting. FFR values are indicated in every test case on the plot

Discussion

High accuracy pressure measurements that were utilized in this study possess the potential to improve the functional assessment of coronary lesions. Not only do these local measurements appear to be more sensitive to flow than FFR, they can also eliminate some of the factors governing the FFR variability since the measurements are sufficiently sensitive to extract information from non-constricted regions as well (e.g., flow rate). Moreover, functional assessment of a lesion based on the local pressure drop, is not dependent on local blood pressure, but only on the absolute pressure drop. In addition to functional assessment of stenoses, it was shown that the fluid pressure drop contains information on the tube distensibility, or compliance. It was also shown that by using very accurate differential pressure measurements, the pressure drop can be measured over smaller segments of the compliant tube (e.g., 1–3 cm), making the local measurement lesion specific.

References

Mokhtari-Dizaji M, Montazeri M, Saberi H (2006) Differentiation of mild and severe stenosis with motion estimation in ultrasound images. Ultrasound Med Biol 32(10):1493–1498. doi:10.1016/j.ultrasmedbio.2006.05.023 (S0301-5629(06)01636-X [pii])

Nakatani S, Yamagishi M, Tamai J, Goto Y, Umeno T, Kawaguchi A, Yutani C, Miyatake K (1995) Assessment of coronary artery distensibility by intravascular ultrasound. Application of simultaneous measurements of luminal area and pressure. Circulation 91(12):2904–2910

Rotman OM, Zaretsky U, Shitzer A, Einav S (2014) Method for high accuracy differential pressure measurements using fluid-filled catheters. Ann Biomed Eng 42(8):1705–1716. doi:10.1007/s10439-014-1026-4

Rotman OM, Weiss D, Zaretsky U, Shitzer A, Einav S (2015) High accuracy differential pressure measurements using fluid-filled catheters—a feasibility study in compliant tubes. J Biomech 48(12):3543–3548. doi:10.1016/j.jbiomech.2015.05.026

Rotman OM, Zaretsky U, Shitzer A, Einav S (2016) Pressure drop and arterial compliance—two arterial parameters in one measurement. J Biomech. doi:10.1016/j.jbiomech.2016.11.026

Tobis J, Azarbal B, Slavin L (2007) Assessment of intermediate severity coronary lesions in the catheterization laboratory. J Am Coll Cardiol 49(8):839–848. doi:10.1016/j.jacc.2006.10.055 (doi:S0735-1097(06)03005-1 [pii])

van Popele NM, Grobbee DE, Bots ML, Asmar R, Topouchian J, Reneman RS, Hoeks AP, van der Kuip DA, Hofman A, Witteman JC (2001) Association between arterial stiffness and atherosclerosis: the Rotterdam Study. Stroke 32(2):454–460

Biomechanical Gait Assessment on a Patient Undergoing Surgical Correction of Kyphosis from Severe Ankylosing Spondylitis: A Case Study

R. Haddas and T. Belanger

Abstract *Background* Ankylosing spondylitis (AS) is a chronic inflammatory disease primarily affecting the axial skeleton, including the sacroiliac joints, costovertebral joints, and the spine. Patients with AS found to have an alter gait pattern. The purpose of this study was to investigate biomechanical alterations in gait after surgical correction in a patient with severe kyphosis from AS. *Methods* A case report in controlled laboratory study, a pretest–posttest design. A 20-year-old male presented with severe sagittal imbalance and inability to stand erect due to AS. He presented with thoracic kyphosis of 70°, lumbar kyphosis of 25°, and pelvic incidence of 43°. The patient had a complex spinal reconstruction with 84° of sagittal correction, normalizing his sagittal alignment. Gait analysis was performed the day before surgery and 1 month post surgery, including three-dimensional kinematics, ground reaction forces, and electromyography outcomes. *Results* Normalization of spinal alignment minimally increased walking speed and cadence. Lower extremity ranges of motion angles increased but were not symmetrical even 1 month post surgery. Postoperatively, trunk flexion, neck extension, and head orientation angles decreased compared with preoperative values but was not symmetrical even 1 month post surgery. The trunk muscles were activated earlier in the post-surgery condition compared to the pre-surgery condition while lower extremity muscles presented later muscle activation. *Conclusions* Surgical correction of spinal alignment improved spine function and efficiency. Changes in gait abnormality parameters observed imply that the patient used less energy to ambulate after surgery than before surgery. Although pre-surgery data showed compensation in the spine kinematics, post-surgery data supported significant changes in the spine and the lower extremity values.

R. Haddas (✉)
Texas Back Institute Research Foundation, Plano, TX, USA
e-mail: rhaddas@texasback.com

T. Belanger
Texas Back Institute, Rockwall, TX, USA

A. Gefen and D. Weihs (eds.), *Computer Methods in Biomechanics and Biomedical Engineering*, Lecture Notes in Bioengineering,
DOI 10.1007/978-3-319-59764-5_23

Introduction

Ankylosing spondylitis (AS) is a chronic inflammatory disease primarily affecting the axial skeleton, including the sacroiliac joints, costovertebral joints, and the spine (Elalouf and Elkayam 2015). The main symptom is inflammatory low-back pain, which is characterized by rest pain accompanied by morning stiffness (Elalouf and Elkayam 2015). In advanced cases, patients present with spinal stiffness, kyphosis with sagittal malalignment, and loss of rib cage motion. Ankylosing spondylitis patients have increased rigidity of the spine, which results in decreased shock absorption and consequently a more cautious gait pattern in the absence of clinically and radiologically detectable peripheral joint disease (Zebouni et al. 1992).

Clinical gait analysis is the process by which quantitative information is collected to aid in understanding the etiology of gait abnormalities and in treatment decision-making (Toosizadeh et al. 2015; Nishimura et al. 2015; Mahaudens and Mousny 2010; Stief et al. 2015; Fatone et al. 2015; Haddas et al. 2016a, b) This process is facilitated through the use of technology such as specialized, computer-interfaced video cameras to measure patient motion and force platforms imbedded in a walkway to monitor the forces and torques produced between the ambulatory patient and the ground (Robertson et al. 2013). Objective motor performance measures, especially gait analysis, could improve evaluation of spine disorder surgeries (Toosizadeh et al. 2015; Nishimura et al. 2015; Mahaudens and Mousny 2010; Stief et al. 2015; Haddas et al. 2016a, b).

Patients with AS found to have an alter gait pattern. Zebouni et al. (1992). compared the gait pattern of patients with ankylosing spondylitis to healthy controls. Those AS patients presented less hip and knee range of motion and reduced stride length. Similar results found by Del Din et al. (2011). who presented reduced in gait velocity and stride length, alterations in the sagittal plane at trunk, pelvis, hip, knee, and ankle for AS patients.

Patient History

A 20-year-old male presented with severe AS, resulting in severe sagittal imbalance and inability to stand erect. The patient had a 10-year history of progressive kyphosis and spinal stiffness. The patient had complete spontaneous fusion of his thoracic and lumbar spine, but had cervical spine mobility (Fig. 1). The patient had complete spontaneous fusion of his thoracic and lumbar spine. He had cervical spine mobility. His deformity was associated with back pain and difficulty with horizontal gaze. He had no history of other medical problems, injury, tuberculosis, or surgery. On physical examination, he had severe kyphosis of the thoracic and lumbar spine, and was neurologically intact. He had negligible coronal imbalance. He also had moderate hip osteoarthritis with hip and knee flexion contractures. Pulmonary function testing revealed significant restrictive lung disease with

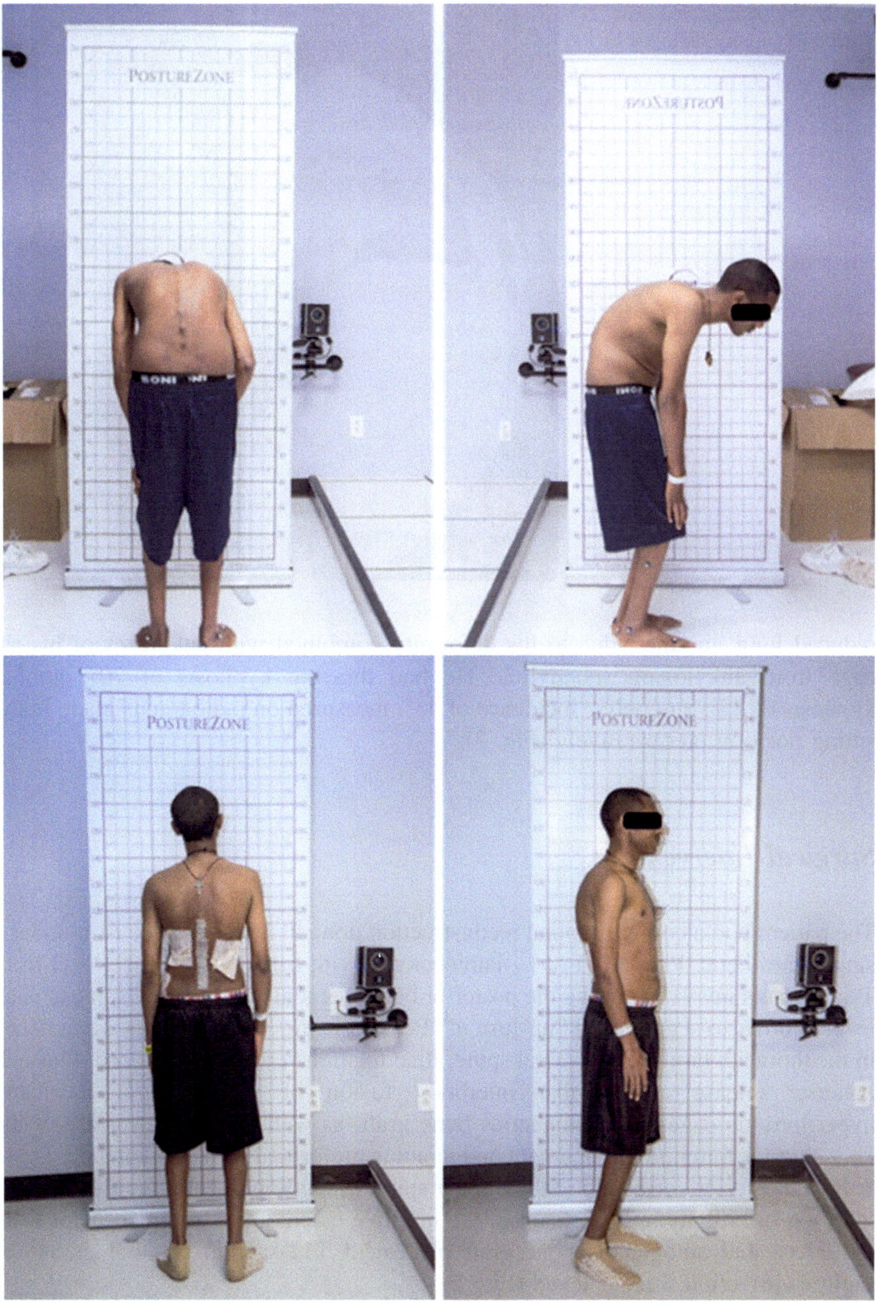

Fig. 1 Pre and post surgery posture

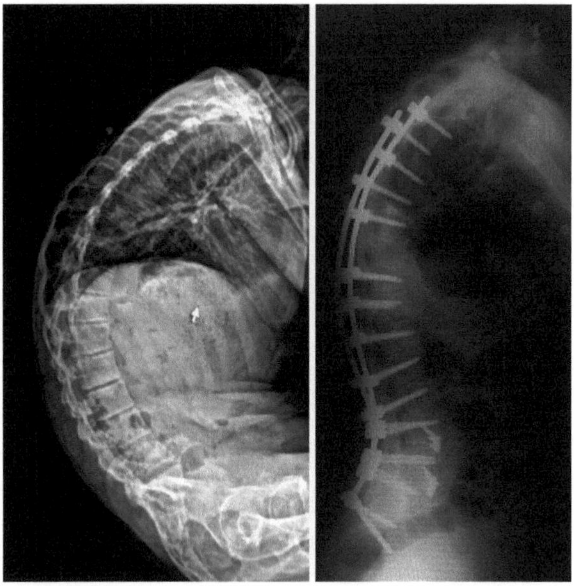

Fig. 2 Pre (sitting) and post (standing) surgery radiographs

reduced lung volumes, due to his deformity combined with ankyloses of his rib cage from ankylosing spondylitis. He had thoracic kyphosis of 70°, lumbar kyphosis of 25°, and pelvic incidence of 43°, measured on radiographs made in the sitting position preoperatively (Fig. 2).

Surgical Intervention

The patient had a complex spinal reconstruction done in 3 phases over 14 h under a single anesthetic. Phase 1: T_{11}–L_5 laminectomy with osteotomy at each level from T_{11}–S_1. This involved cutting the posterior bony elements of the fused spine at each level in order to segmentally "un-fuse it". Pedicle screw fixation points were placed in the thoracic, lumbar, and sacral spine. Iliac fixation screws were placed. Phase 2: Anterior release and lumbar interbody fusion of $L_{4–5}$ and L_5–S_1 using hyper-lordotic cages and autogenous bone graft, as well as anterior release of the $L_{3–4}$ anterior longitudinal ligament and annulus preformed from the left side. Phase 3: Final posterior correction of the lumbar spine with completion of the $L_{3–4}$ Smith–Peterson osteotomy, compression of the multiple osteotomies, rod placement from T_5-Pelvis, and autogenous bone graft placement T11-sacrum. A total of 84° of sagittal correction was achieved (Fig. 2).

The effect of complex spinal reconstruction on AS patients (de Vries et al. 2010; Etame et al. 2008; Kubiak et al. 2005; Liu et al. 2015) and their gait pattern (Zebouni et al. 1992; Fatone et al. 2015; De Filippis et al. 2006) are well documented. However, there is limited literature on whether complex spinal reconstruction effects on spine and lower extremity biomechanics during gait in patients with AS. The purpose of this study was to investigate biomechanical alterations in gait after surgical correction in a patient with severe kyphosis from AS.

Materials and Methods

A case report in controlled laboratory study, a pretest–posttest design was used in this study. Gait analysis was performed the day before surgery and 1 month post surgery. Fifty-one reflective markers (9.5 mm diameter) were incorporated to collect full body three-dimensional kinematics using 10 cameras (VICON, Denver, CO) at a sampling rate of 100 Hz. A static trial was then collected to note marker placement. Ground reaction forces (GRFs) were measured at 2000 Hz using three parallel force plates (AMTI, Watertown, MA). The patient walked barefoot at his self-selected speed along a 10 m walkway. At least 5 trials were recorded during each session. Scaled and transformed 3D motion capture marker data, and scaled GRF data were exported from the Vicon Nexus system and imported into a custom MATLAB (Mathworks Inc., Natick, MA) program for processing. Marker coordinate data were smoothed using a fourth-order no-phase-shift low pass Butterworth digital filter (6 Hz cutoff) and used to calculate joint angle time series data for the hip, pelvis, spine segments, and head (Haddas et al. 2016b). Gait cycle define as the time period of movements in which one foot contacts the ground to when that same foot again contacts the ground, and involves forward propulsion of the centre of gravity. The ground reaction force (GRF) data were not filtered during post-acquisition processing. Clinical gait analysis parameters were calculated from kinematic and kinetic data: spatiotemporal parameters, joint angle values at a specific instant of the gait cycle, and peak GRF.

Results

Overall, surgical correction of severe kyphosis improved the patient's gait performance and efficiency (Fig. 3). Normalization of spinal alignment minimally increased walking speed (0.02 m/s) and cadence (4 steps/min), but did not change step lengths. Trunk flexion decreased (63°) along with neck extension (19°) and head orientation (43°) angles at initial contact. Lower extremity range of motion (ROM) increased, but was not symmetrical even 1 month post surgery. Bilateral hip (R 10°; L 2°), left knee (9°), and bilateral ankle (R 5°; L 30°) ROM increased during the stance phase while right knee (11°) ROM decreased compared to the pre surgery condition (Figs. 3 and 4). Presurgery data showed, the majority of body compensation for the tremendous trunk flexion came from the neck and head and less from the lower extremities. Ground reaction forces (GRFs) were decreased (R 21N; 29N) post surgery.

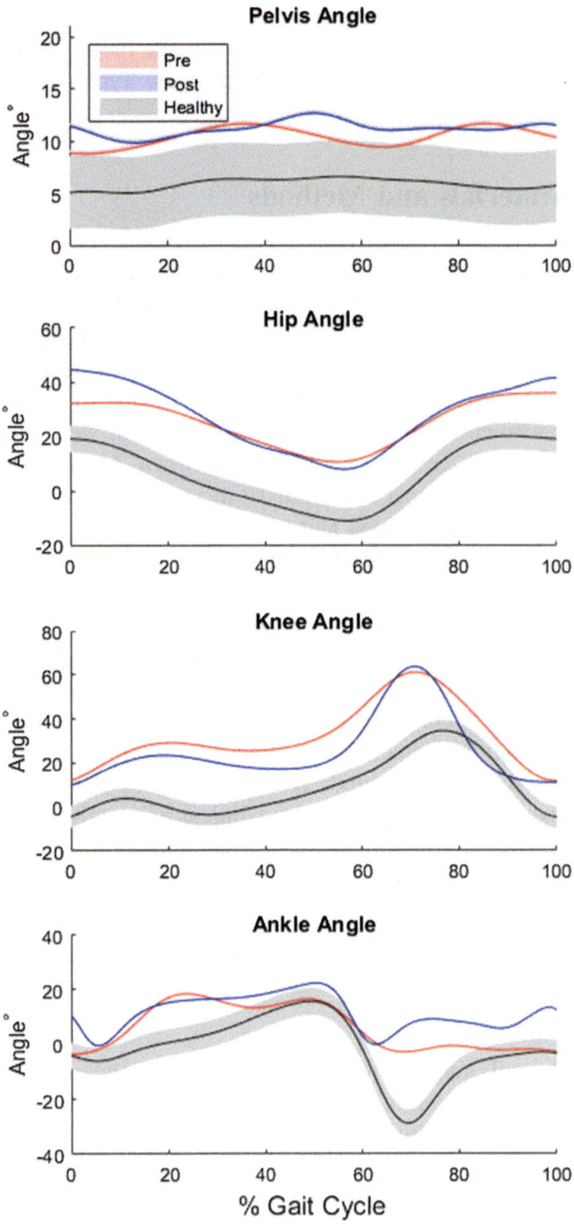

Fig. 3 Sagittal plane excursions of the pelvis, hip, knee and ankle pre- (*red*) and post- (*blue*) surgical spinal alignment for a patients with severe ankylosing spondylitis in comparison the healthy values (*gray*)

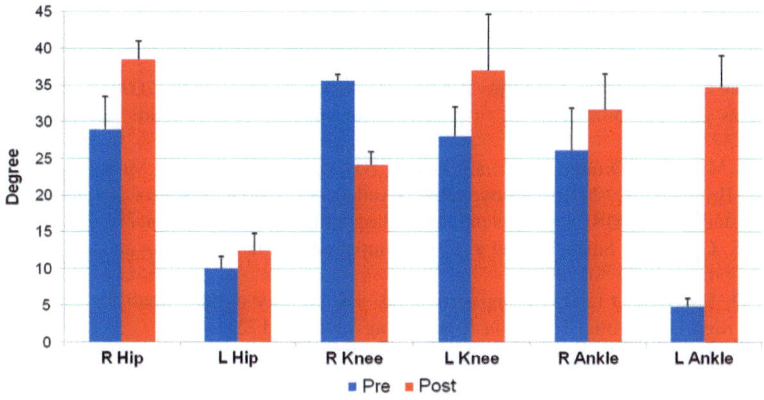

Fig. 4 Lower extremity range of motion sagittal plane during the stance phase. Positive flexion/plantarflexion. *Pre* 1 Day before Surgery, *Post* 1 Month after Surgery

Discussion

The purpose of this study was to investigate biomechanical alterations in gait after surgical correction in a patient with severe kyphosis from AS. Overall, surgical correction of spinal alignment improved lower extremity function. Although gait parameters did not significantly improve, trunk and lower extremity kinematic dramatically improve due to surgical correction. Our results Clinical gait analysis reveals significant difference before and after spine surgery when impending to a normal gait values (Fig. 3). The patient appeared to leaning slightly to the left and flexed while demonstrating more right knee flexion contracture than the left pre-operative. To maintain this posture preoperatively, he may be ranging his left knee less than the right. The right knee ROM may have decreased postoperatively due to the correction in the coronal plane. After the coronal correction, his left knee would require a greater degree of extension to maintain the balance while the right knee would require less range of motion to maintain this stance. This may explain the increase in the right ankle ROM to compensate for the less ROM of the right knee and the increased ROM of the left ankle to compensate for the persistent decreased ROM of the left hip, most likely due to the hip flexion contracture. Left side approach was used for the anterior release; this may explain the postoperative decrease in the left IO and EO.

Changes in gait parameters observed imply that the patient used less energy to ambulate after surgery than before surgery (VanSwearingen and Studenski 2014) based on the joint kinematics and GRF data. The decrease in GRF also is the result of more efficient gait and increased speed.

References

De Filippis LG, Balestrieri A, Furfari P, Caliri A, Africa A, Bagnato G (2006) Muscle activation patterns and gait biomechanics in patients with ankylosing spondylitis. Reumatismo 58(2): 132–137

de Vries MK, van Drumpt AS, van Royen BJ, van Denderen JC, Manoliu RA, van der Horst-Bruinsma IE (2010) Discovertebral (Andersson) lesions in severe ankylosing spondylitis: a study using MRI and conventional radiography. Clin Rheumatol 29(12):1433–1438

Del Din S, Carraro E, Sawacha Z et al (2011) Impaired gait in ankylosing spondylitis. Med Biol Eng Comput 49(7):801–809

Elalouf O, Elkayam O (2015) Long-term safety and efficacy of infliximab for the treatment of ankylosing spondylitis. Ther Clin Risk Manag 11:1719–1726

Etame AB, Than KD, Wang AC, La Marca F, Park P (2008) Surgical management of symptomatic cervical or cervicothoracic kyphosis due to ankylosing spondylitis. Spine 33(16):E559–E564

Fatone S, Stine R, Gottipati P, Dillon M (2015) Pelvic and spinal motion during walking in persons with transfemoral amputation with and without low back pain. Am J Phys Med Rehabil

Haddas R, Sawyer SF, Sizer PS, Brooks T, Chyu MC, James CR (2016a) Effects of volitional spine stabilization and lower extremity fatigue on trunk control during landing in a recurrent low back pain population. J Orthop Sports Phys Ther 1:1–23

Haddas R, Yang J, Lieberman I (2016b) Effects of volitional spine stabilization on lifting task in recurrent low back pain population. Eur Spine J 25(9):2833–2841

Kubiak EN, Moskovich R, Errico TJ, Di Cesare PE (2005) Orthopaedic management of ankylosing spondylitis. J Am Acad Orthop Surg 13(4):267–278

Liu H, Yang C, Zheng Z et al (2015) Comparison of smith-petersen osteotomy and pedicle subtraction osteotomy for the correction of thoracolumbar kyphotic deformity in ankylosing spondylitis: a systematic review and meta-analysis. Spine 40(8):570–579

Mahaudens M, Mousny M (2010) Gait in adolescent idiopathic scoliosis. Kinematics, electromyographic and energy cost analysis. In: Carl-Eric Aubin IAFS, Hubert L, Alain M (ed) Research into spinal deformities, vol. 7, p 158

Nishimura H, Endo K, Suzuki H, Tanaka H, Shishido T, Yamamoto K (2015) Gait analysis in cervical spondylotic myelopathy. Asian Spine J 9(3):321–326

Robertson GE, Caldwell GE, Hamill J, Kamen G, Whittlesey SN (2013) Research methods in biomechanics. Human Kinetics, Champaign

Stief F, Meurer A, Wienand J, Rauschmann M, Rickert M (2015) Has a mono- or bisegmental lumbar spinal fusion surgery an influence on self-assessed quality of life, trunk range of motion, and gait performance? Spine (Phila Pa 1976) 40(11):E618–E626

Toosizadeh N, Yen TC, Howe C, Dohm M, Mohler J, Najafi B (2015) Gait behaviors as an objective surgical outcome in low back disorders: a systematic review. Clin Biomech 30(6): 528–536

VanSwearingen JM, Studenski SA (2014) Aging, motor skill, and the energy cost of walking: implications for the prevention and treatment of mobility decline in older persons. J Gerontol A Biol Sci Med Sci 69(11):1429–1436

Zebouni L, Helliwell PS, Howe A, Wright V (1992) Gait analysis in ankylosing spondylitis. Ann Rheum Dis 51(7):898–899

Walking Sticks and a Walker Comparison during Gait in Adult Degenerative Scoliosis Patients

Ram Haddas and Isador Lieberman

Abstract Adult degenerative scoliosis results from age-related changes leading to segmental instability, deformity, and stenosis. Patients with degenerative adult scoliosis demonstrate an altered gait pattern. Use of a walker tends to cause a more kyphotic posture due to its lower hand grips, and due to the way patients must reach forward with this position using both hands, forcing a kyphotic moment into their gate cycle. Whereas a walker forces patients into kyphosis, the higher grips of walking sticks allow for more upright posture and improved sagittal alignment. The purpose of this study is to compare and contrast the benefits of walking sticks versus a walker on the biomechanics of the lower extremity in people with degenerative scoliosis, as evaluated by gait analysis. Ten patients with symptomatic degenerative scoliosis have been deemed appropriate surgical candidates. Each patient performed a series of overground gait trials with a comfortable self-selected speed under three testing conditions: 1. with walking sticks, 2. with a walker, and 3. without any device. The use of walking sticks resulted in significantly slower, longer stride and step times along with bigger ankle, knee, and hip flexion range of motions in comparison to the walker. Walking sticks did improve the biomechanics and did facilitate positioning of the trunk and lower extremity during walking in adult degenerative scoliosis patients.

Introduction

Adult degenerative scoliosis results from age-related changes leading to segmental instability, deformity, and stenosis. Although the etiology is unclear, degenerative adult scoliosis is associated with progressive and asymmetric degeneration of the disc and facet joints, which typically lead to stenosis (Kotwal et al. 2011). By virtue

R. Haddas (✉)
Texas Back Institute Research Foundation, Plano, TX, USA
e-mail: rhaddas@texasback.com

I. Lieberman
Texas Back Institute, Plano, TX, USA

A. Gefen and D. Weihs (eds.), *Computer Methods in Biomechanics and Biomedical Engineering*, Lecture Notes in Bioengineering,
DOI 10.1007/978-3-319-59764-5_24

of the narrowed spinal canal associated with the degeneration these patients frequently develop back pain, as well as leg pain, weakness, and numbness (Kotwicki et al. 2013). With an aging population in the USA and an increased attention to quality of life versus cost issues in the current healthcare environment, degenerative adult scoliosis has become a considerable healthcare concern (Kotwal et al. 2011; Ploumis et al. 2007). Patients with degenerative adult scoliosis demonstrate an altered gait pattern (Yang et al. 2013). Such differences include decreased step length and reduced range of motion in the upper and lower extremities (Mahaudens et al. 2009), asymmetry of trunk rotation (Quervain et al. 2004), and ground reaction force (Quervain et al. 2004; Schizas et al. 1998; Chockalingam et al. 2004; Giakas et al. 1996) in three dimensions. Mahaudens et al (2009) found a decrease in the muscular mechanical work associated with an increase of energy cost and a decrease in the muscular efficiency in degenerative adult scoliosis population. Furthermore, degenerative adult scoliosis patients exert 30% more physical effort than healthy subjects to ensure habitual locomotion, and this additional effort requires a reciprocal increase of oxygen consumption (Mahaudens et al. 2009). This altered gait pattern demonstrated by subjects with degenerative adult scoliosis may be due to changes in global postural control strategies caused by spinal deformity (Yang et al. 2013).

Rollator walkers are frequently prescribed in an effort to improve balance, mobility, and reduce lower back pain. Clinical experience has shown the use of walking sticks rather than a walker promotes a more upright posture. Alkjaer et al (2006) analyzed the 3D walking pattern during walking with and without walkers. Their results demonstrated walking with a walker unloaded the ankle and knee extensors, while increasing the hip flexion angle. Moreover, the use of the walker did not result in an overall unloading of the muscles and joints of the lower extremities (Alkjær et al. 2006). Youdas et al (2005) found a reduction in the vertical ground reaction force during the use of a rollator walker. Bryant et al. (2012), on the other hand, found that walkers did not affect cadence, double support phase, heel to heel base of support, stride time, and stance period.

More recently, walking sticks have been prescribed in the preoperative and postoperative periods to assist with mobility, promote a more balanced upright posture and trigger activation of the spine stabilizer muscles during the gait cycle. Whereas a walker forces patients into kyphosis, the higher grips of walking sticks allows for more upright posture and improved sagittal alignment. Studies on gait with walking sticks in healthy individuals have shown that these walking-aids reduce the load on the lower extremities (Willson et al. 2001). Usage of walking sticks increased walking speed, stride length, and stance time, decrease in vertical and anterior–posterior ground reaction forces (GRF), and a decrease in compressive knee joint reaction force compared with the no walking sticks condition (Willson et al. 2001). On the other hand, Bechard et al. (2012) did not find any changes in knee function while using walking sticks and recommend not to use it to decrease knee joint loading. Furthermore, there were no differences in the change in step length, forward velocity, or cadence between the group that used walking sticks and

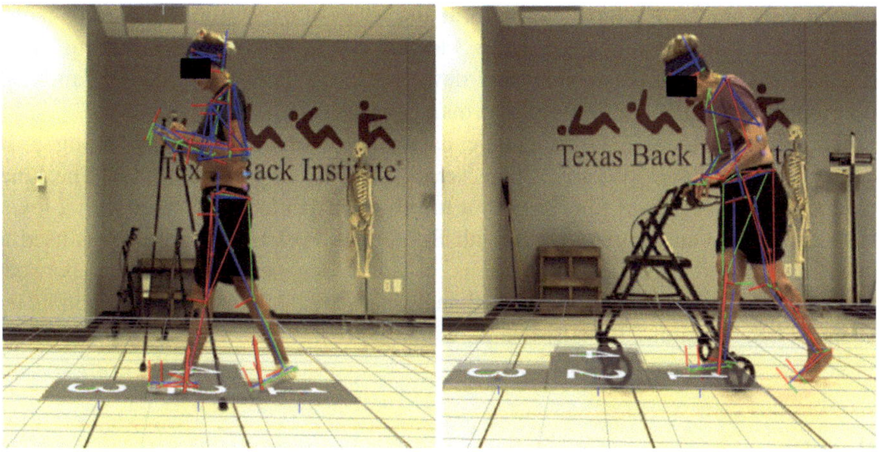

Fig. 1 Gait with walking sticks and with walker in adult degenerative scoliosis patients

the standard walking exercise-training program group at baseline and the 6-week measurement period (Collins et al. 2012).

Based on our knowledge, there is limited literature on biomechanics and neuromuscular control of the lower extremities and spine as assessed by objective gait analysis with walking sticks versus a walker, in adult degenerative scoliosis patients, pre and post surgical intervention. The purpose of this study is to compare and contrast the benefits of walking sticks versus a walker on the biomechanics of the lower extremity in people with degenerative scoliosis, as evaluated by gait analysis using video motion capture and force plate analysis.

Materials and Methods

Ten patients (Age: 68 ± 6.4; Weight: 69.8 ± 3.64 kg; Height: 1.63 ± 0.16 m) with symptomatic degenerative scoliosis who have been deemed appropriate surgical candidates. Patients were included in the study if they were able to ambulate without assistance, and consented to the gait analysis. Patients were excluded if they had history of prior spine or major lower extremity surgery or previous injury that may affect gait, BMI higher than 35, primary neurological disorder other than cervical myelopathy, diabetic neuropathy or other disease that impairs the patient's ability to ambulate or stand without assistance, and pregnant. Healthy volunteers were recruited from the general population.

All test patients were fitted with 51 external reflective markers using locations and procedures modified from Vaughan et al. (1999). These markers were placed on the skin overlying strategic anatomic points. A static video trial was recorded with subjects positioned in a neutral, standing posture to create a reference for defining neutral joint angles. Each patient performed a series of over-ground gait trials with a comfortable self-selected speed under 3 testing conditions; (1) with walking sticks, (2) with walker, and (3) without any device (Fig. 1). Patients walked 10 m, stepping on three sequenced force platforms.

Three-dimensional (3D) kinematic data was recorded at 200 Hz using a 10 camera Vicon Video system (Vicon Nexus 2.0 Inc., Englewood, CO). The kinematic data was low pass filtered with a fourth-order Butterworth filter with a lower cutoff at 4 Hz. One-way ANOVA with Bonferroni Post Hoc analyses was used to determine differences in gait patterns in adult degenerative scoliosis between walking devices following by post-hoc analysis. Statistical analyses were conducted using SPSS, Version 23.0 (IBM, Inc., Chicago, IL).

Results

The use of walking sticks resulted in significant slower cadence (22.8 step/min $p < 0.00$ compared to no device and 6.8 step/min $p < 0.02$). Stride and step times were longer compared to the walker (0.36 s $p < 0.00$ and 0.52 s $p < 0.01$, respectively) and with no device (0.68 s $p < 0.00$ and 0.39 s $p < 0.01$, respectively). Walking sticks significantly resulted in more dorsiflexion angle at initial contact (Right 6.9° $p < 0.02$; Left 6.4° $p < 0.03$) in comparison to the walker. Head orientation presented more extension position using the walking sticks (6.7° $p < 0.05$) and the walker (8.6° $p < 0.04$). The use of the walking sticks significantly increase ankle (21.6° $p < 0.00$), knee (24.6° $p < 0.01$), and hip (15.0° $p < 0.00$) flexion range of motions in comparison to the walker. Moreover, the walking sticks promoted more knee (15.6° $p < 0.00$), hip (3.9° $p < 0.00$), and head (4.0° $p < 0.00$) range of motion in the frontal plane in comparison to a walker.

Conclusion

This study proved that simply changing the type of walking aid from a walker to walking sticks, quantifiable evidence of improved gait cycle parameters can be obtained. Both the slower cadence and walking speed that was found in the walking stick group was attributed to unfamiliarity with the devices. The best length of walking stick has not been determined either, and is currently set at the patient's shoulder. This could change as we gain a better understanding of gait with these devices.

Walking sticks did improve the biomechanics and did facilitate positioning of the trunk and lower extremity during walking in adult degenerative scoliosis patients. With preoperative walking stick training, surgical correction of deformity, and postoperative use of walking sticks, improvement in both sagittal and frontal parameters and kinematics as compared to a walker can be expected. We recommend the use of walking sticks to assist with ambulation in symptomatic adult degenerative scoliosis patients.

Formal gait and motion analysis can provide a method to assess the impact of severe spinal deformity on function and changes after treatment. Moreover, gait analysis for ADS is a new area of research with potential benefits that include a better understanding of patient outcomes as they relate to sagittal and coronal alignment, dynamic parameters such as walking speeds and step times, and overall kinetic and kinematic parameters that shed light on how the patient is moving each joint in their body.

References

Alkjær T, Larsen PK, Pedersen G, Nielsen LH, Simonsen EB (2006) Biomechanical analysis of rollator walking. Biomed Eng Online 5:2

Bechard DJ, Birmingham TB, Zecevic AA et al (2012) The effect of walking poles on the knee adduction moment in patients with varus gonarthrosis. Osteoarthr Cartil 20(20):12

Bryant MS, Pourmoghaddam A, Thrasher A (2012) Gait changes with walking devices in persons with parkinson's disease. Disabil Rehabil Assist Technol 7(2):149–152

Chockalingam N, Dangerfield PH, Rahmatalla A, Ahmed N, Cochrane T (2004) Assessment of ground reaction force during scoliotic gait. Eur Spine J 13(8):750–754

Collins EG, O'Connell S, McBurney C et al (2012) Comparison of walking with poles and traditional walking for peripheral arterial disease rehabilitation. J Cardiopulm Rehabil Prev 32 (4):210–218

Giakas G, Baltzopoulos V, Dangerfield PH, Dorgan JC, Dalmira S (1996) Comparison of gait patterns between healthy and scoliotic patients using time and frequency domain analysis of ground reaction forces. Spine (Phila Pa 1976) 21:2235–2242

Kotwal S, Pumberger M, Hughes A, Girardi F (2011) Degenerative scoliosis: a review. HSS J 7 (3):257–264

Kotwicki T, Chowanska J, Kinel E, Czaprowski D, Tomaszewski M, Janusz P (2013) Optimal management of idiopathic scoliosis in adolescence. Adolesc Health Med Ther 4:59–73

Mahaudens P, Banse X, Mousny M, Detrembleur C (2009a) Gait in adolescent idiopathic scoliosis: kinematics and electromyographic analysis. Eur Spine J 18:512–521

Mahaudens P, Detrembleur C, Mousny M, Banse X (2009b) Gait in adolescent idiopathic scoliosis: energy cost analysis. Eur Spine J 18(8):1160–1168

Ploumis A, Transfledt EE, Denis F (2007) Degenerative lumbar scoliosis associated with spinal stenosis. Spine J 7(4):428–436

Quervain K-D, Muller R, Stacoff A, Grob D, Stussi E (2004) Gait analysis in patients with idiopathic scoliosis. Eur Spine J 13:449–456

Schizas CG, Quervain K, Stussi E, Grob D (1998) Gait asymmetries in patients with idiopathic scoliosis using vertical forces measurement only. Eur Spine J 7:95–98

Vaughan CL, Davis BL, O'Conner JC (1999) Dynamics of human Gait. Kiboho Publishers, Cape Town

Willson J, Torry MR, Decker MJ, Kernozek T, Steadman JR (2001) Effects of walking poles on lower extremity gait mechanics. Med Sci Sports Exerc 33(1):142–147

Yang JH, Suh S-W, Sung PS, Park W-H (2013) Asymmetrical gait in adolescents with idiopathic scoliosis. Eur Spine J 22(11):2407–2413

Youdas JW, Kotajarvi BJ, Padgett DJ, Kaufman KR (2005) Partial weight-bearing gait using conventional assistive devices. Arch Phys Med Rehabil 86(3):394–398

Finite Element Analysis of Pre and Post Lumbar Fusion for Adult Degenerative Scoliosis Patients

Ram Haddas, Ming Xu, Isador Lieberman and James Yang

Abstract *Purpose*: To investigate the effect of adjacent load transfer pre- and post-fusion surgery of lumbar scoliotic spines using FE models. *Methods*: Ten three-dimensional nonlinear FE models of the lumbosacral spine were created from pre- (Cobb angle: 28.1° ± 10.5°) and post-scoliosis surgery and healthy in vivo CT scans. During surgery, pedicle screws and rods were implanted at lumbar and sacral levels. A compressive load and six different moments (flexion, extension, lateral bending, and axial rotation) were applied to the top level of each model. Outcome measures were range of motion (RoM), intradiscal pressure (IDP), and facet joint forces (FJF). *Results*: Spinal fusion did alter the mechanical function of the scoliotic spine. Scoliotic spine presented abnormal and asymmetrical kinetic and kinematic behavior in comparison to a healthy spine. After the fusion surgery, RoM decreased by 91.88 %, IDP decreased by 46.87 %, and FJF decreased by 60.63 % at the fused level on average whereas a minor increase of RoM, IDP, or FJF was observed at the adjacent level. Compared to the healthy subjects, the pre-surgical scoliosis subjects have up to 8.03 % greater RoM, 20.04 % increased IDP, and 18.38 % increased FJF on average at the adjacent level. *Conclusions*: This study was the first to investigate the effect of adjacent load transfer before and after fusion surgery using in vivo CT scans of ten scoliotic spines. A posterior fusion has only a minor effect on mechanical behavior and large effect on pressure and forces at the adjacent level. As expected, a large effect on the kinematics and kinetics was found at the fused level.

R. Haddas (✉)
Texas Back Institute Research Foundation, Plano, TX, USA
e-mail: rhaddas@texasback.com

M. Xu · J. Yang
Mechanical Engineering, Texas Tech University, Lubbock, TX, USA

I. Lieberman
Texas Back Institute, Plano, TX, USA

A. Gefen and D. Weihs (eds.), *Computer Methods in Biomechanics and Biomedical Engineering*, Lecture Notes in Bioengineering,
DOI 10.1007/978-3-319-59764-5_25

Introduction

Adult degenerative scoliosis (ADS) results from age-related changes leading to segmental instability, deformity, and stenosis (Kotwal et al. 2011). Although the etiology is unclear, degenerative adult scoliosis is associated with progressive and asymmetric degeneration of the disc and facet joints, which typically leads to stenosis (Kotwal et al. 2011). By virtue of the narrowed spinal canal associated with the degeneration, these patients frequently develop back pain, as well as leg pain, weakness, and numbness (Kotwicki et al. 2013). Spinal fusion is a standard surgical treatment for deformity (Barrey and Darnis 2015). This surgery has proven to be successful when performed for the correct indication. The surgery may involve decompression, instrumentation, and fusion to stabilize the spine and alleviate the symptoms.

In comparison to in vitro or in vivo approaches, finite element (FE) studies eliminate the issues with cadaver use in addition to being cost efficient, time efficient, and an accurate biomechanical surrogate (Dreischarf et al. 2014; Bloemker et al. 2012). FE study is also capable of capturing the biomechanical parameters internal to the bones and connective soft tissues of the spine, which is difficult to measure by experimental approaches (Little et al. 2008). Several studies have investigated a single scoliosis spine using FE study pre- or post-surgery only (Zheng et al. 2015; Wang et al. 2016). Furthermore, previous studies in the literature (Ambati et al. 2015; Rohlmann et al. 2007) studied the biomechanics of the scoliotic spines using one deterministic healthy spine or a modified healthy spine, which was not able to fully represent the anatomical difference between the healthy spines and scoliotic spines.

Due to the inter-subject variations among the spines especially scoliotic spines where larger inter-subject variations are expected compared to those in the healthy spines, the predictions obtained by one deterministic FE model are insufficient and not generalizable (Dreischarf et al. 2014). Literature that describe the underlying mechanisms responsible for spinal fusion in scoliosis patients is limited and FE study with larger subject sample size should be conducted (Dreischarf et al. 2014). The purpose of this study was to investigate the effect of adjacent load transfer before and after fusion surgery of lumbar scoliotic spines using the FE models from in vivo CT scans of ten scoliotic spines based on the validated FE modeling method in healthy subjects (Xu et al. 2016).

Materials and Methods

Five patients (age: 60.9 ± 15.0, Table 1) pre- and post-scoliosis surgery CT scans were used to produced 10 three-dimensional (3D) nonlinear FE models of the scoliotic lumbosacral spine. The lumbosacral spine FE model was developed and validated in our previous work (Xu et al. 2016). Subject information including age,

Table 1 Ten scoliosis FE models tested in the study

Age	Gender	Segment included	Fused segments	FE pre surgery	FE post surgery
81	Male	T12–S1	L1–S1	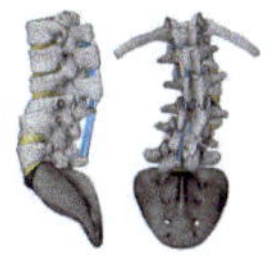	
64	Male	T11–S1	T12–L5	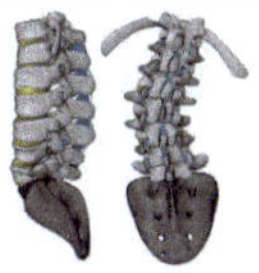	
57	Male	T12–S1	L3–S1	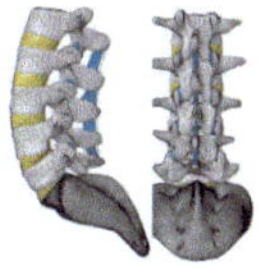	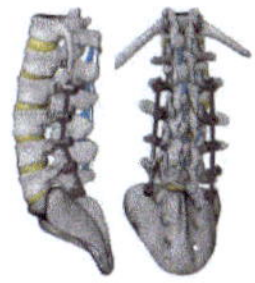
39	Male	L1–S1	L2–L4	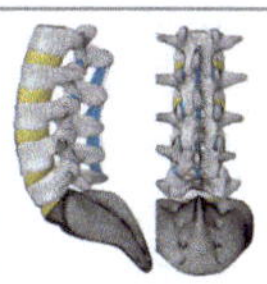	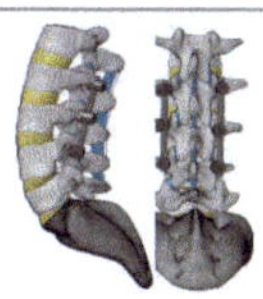
30	Male	L2–S1	L5–S1	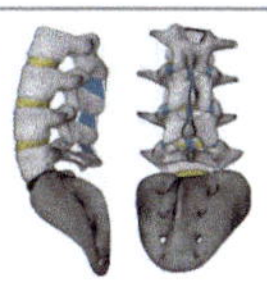	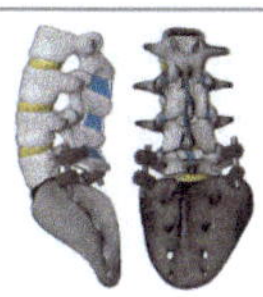

gender, and fused spinal segments were listed in Table 1. For each subject, one pre-surgical and one post-surgical FE model were developed (Table 1). Furthermore, five three-dimensional nonlinear FE models of the healthy lumbosacral spine were created based on CT scans of five healthy subjects (Xu et al. 2016). The two-dimensional (2D) CT images were imported into 3D Slicer® (Fedorov et al. 2012) to generate 3D geometries of the vertebrae. Then, a smoothing process was performed in Geomagic Studio® (3D Systems, Rock Hill, SC, USA) to remove spikes and holes on the surface of the vertebral geometries. The geometries of intervertebral discs were created in Geomagic Studio based on the adjacent vertebrae. All seven major ligaments were meshed by 4-node shell elements (Bowden et al. 2008). Local muscle forces and upper body weight in lumbar spine were simulated by a compressive follower load with optimized path through the vertebrae (Dreischarf et al. 2010). In this study, all five post-surgical scoliosis subjects were fused with bilateral pedicle screw fixation systems. The pedicle screws and rods were modeled as linear-elastic and homogeneous Titanium. The material properties assigned for the FE models in this study are all validated (Xu et al. 2016).

The loading process involved two steps. First, a discrete beam (optimized path) was generated and a preload was applied as a follower load along this path (Xu et al. 2016; Dreischarf et al. 2010) where the preload represented the upper body weight and muscle forces. To minimize the intervertebral rotation and shear force introduced by the preload, the preload was adjusted to pass through the instantaneous centers of rotation at each individual level (Dreischarf et al. 2010). The intervertebral rotation caused by the preload should be less than 0.5°, which was considered to be negligible (Schmidt et al. 2012). After the FE models stabilized from the preload, the state was saved and used as an initial condition for the second step. In the second step, further loadings of rational moments were applied for all six lumbar spine movement directions: flexion (+)/extension (−), right (+)/left (−) lateral bending and right (+)/left (−) axial rotation respectively. The loading conditions were applied at the superior endplate of the upmost vertebral body. The sacrum along with the interior side of the L_5/S_1 intervertebral disc in each model was rigidly constrained for all degrees of freedom.

The loading modes tested in this study were taken from the literature and proved to represent realistic maximal voluntary motions as measured in vivo (Dreischarf et al. 2014; Dreischarf et al. 2011). Under each of the six loading conditions, range of motion (ROM), intradiscal pressure (IDP), and facet joint force (FJF) were recorded for the upmost fused spinal level and the lowest adjacent spinal level. The simulation results of the scoliosis subjects tested in this study were compared with the simulation results obtained by the healthy subjects subjected to the identical loading modes in our previous study (Xu et al. 2016).

Repeated measurement ANOVA used to determine differences in ROM, IDP, and FJF in scoliosis pre- and post-surgery FE models. Furthermore, One-way ANOVA with Bonferroni Post Hoc analyses was used to determine differences in ROM, IDP, and FJF in scoliosis pre- and post-surgery FE to the healthy models. Statistical analyses were conducted using SPSS, Version 23.0 (IBM, Inc., Chicago, IL).

Results

Range of Motion

At the adjacent level, no significant difference between the pre-surgical and post-surgical scoliosis subjects and between the healthy and pre-surgical scoliosis subjects were observed. At the adjacent level, spinal fusion surgery produced a significant increased left (Pre: 4.18° ± 0.54°, Post: 4.53° ± 0.68°; $p = 0.001$) and right later bending ROM (Pre: 4.01° ± 0.84°, Post: 4.27° ± 0.85°; $p = 0.006$) in comparison to pre-surgical scoliosis models. At the fused level, spinal fusion surgery produced a significantly reduced ROM in all the loading conditions (Flexion Pre: 5.82° ± 0.89°, Post: 0.34° ± 0.31°; $p = 0.001$, Extension Pre: 3.63° ± 0.53°, Post: 0.66° ± 0.31°; $p = 0.001$, Left lateral bending Pre: 4.22° ± 0.61°, Post: 0.16° ± 0.08°; $p = 0.001$, Right lateral bending Pre: 4.30° ± 0.59°, Post: 0.16° ± 0.05°; $p = 0.001$, Left axial rotation Pre: 1.54° ± 0.65°, Post: 0.13° ± 0.19°; $p = 0.001$, Right axial rotation Pre: 1.58° ± 0.76°, Post: 0.14° ± 0.18°; $p = 0.001$) in comparison to pre-surgical scoliosis models. No significant differences were found between the healthy to the pre- and post-surgery models at the adjacent level. In comparison to the healthy models, pre-surgery model presented significantly reduced extension (Pre: 3.63° ± 0.53°, Healthy: 4.34° ± 0.29°; $p = 0.015$) and left later bending (Pre: 4.22° ± 0.61°, Healthy: 5.16° ± 1.19°; $p = 0.042$) ISR in the fused level. Moreover, the post-surgery model presented significantly reduced ISR (Flexion Post: 0.34° ± 0.31°, Healthy: 5.76° ± 1.40°; $p = 0.001$, Extension Post: 0.66° ± 0.31°, Healthy: 4.34° ± 0.29°, $p = 0.\ 001$; Left lateral bending Post: 0.16° ± 0.08°, Healthy: 5.16° ± 1.19°, $p = 0.\ 001$; Right lateral bending Post: 0.16° ± 0.55°, Healthy: 5.16° ± 1.19°, $p = 0.\ 001$; left axial rotation Post: 0.13° ± 0.19°, Healthy: 1.55° ± 0.27°, $p = 0.\ 001$; Right axial rotation Post: 0.14° ± 0.18°, Healthy: 1.55° ± 0.27°, $p = 0.\ 001$) in all the loading conditions in the fused level.

Intradiscal Pressure

At the fused level, spinal fusion surgery produced a much lower IDP in all the loading conditions (Flexion Pre: 1.74 ± 0.24 MPa, Post: 1.18 ± 0.28 MPa; $p = 0.001$, Extension Pre: 0.71 ± 0.20 MPa, Post: 0.29 ± 0.13 MPa; $p = 0.001$, Left lateral bending Pre: 1.03 ± 0.27 MPa, Post: 0.45 ± 0.20 MPa; $p = 0.001$, Right lateral bending Pre: 1.05 ± 0.242 MPa, Post: 0.46 ± 0.19 MPa; $p = 0.001$, Left axial rotation Pre: 0.82 ± 0.25 MPa, Post: 0.49 ± 0.15 MPa; $p = 0.001$; Right axial rotation Pre: 0.83 ± 0.25 MPa, Post: 0.49 ± 0.15 MPa; $p = 0.001$). In comparison to the healthy models, post-surgery model exhibited a reduced IDP in all the loading conditions in the fused level (Flexion Healthy: 1.64 ± 0.27 MPa, Post: 1.18 ± 0.28 MPa; $p = 0.001$, Extension Healthy: 0.54 ± 0.06 MPa, Post: 0.29 ± 0.13 MPa; $p = 0.001$, Left lateral bending Healthy: 0.83 ± 0.23 MPa, Post: 0.45 ± 0.20 MPa; $p = 0.001$, Right lateral bending Healthy: 0.83 ± 0.23 MPa, Post: 0.46 ± 0.19 MPa; $p = 0.001$, Left axial rotation Healthy: 0.71 ± 0.23 MPa, Post: 0.49 ± 0.15 MPa; $p = 0.001$; Right axial rotation Healthy: 0.79 ± 0.23 MPa, Post: 0.49 ± 0.15 MPa; $p = 0.001$). At the adjacent level, no significant differences were found between the healthy to the pre-surgery models in the fused level or between the healthy to the pre- and post-surgery models.

Facet Joint Force

At the adjacent level, spinal fusion surgery produced a considerable larger left lateral rotation FJF (Pre: 80.39 ± 13.13N, Post: 87.95 ± 13.55N; $p = 0.001$) in comparison to pre-surgical scoliosis models. At the fused level, spinal fusion surgery produced considerably lower FJF in all the loading conditions (Extension Pre: 60.24 ± 20.74N, Post: 30.34 ± 12.28N; $p = 0.001$, Left lateral bending Pre: 63.00 ± 18.15N, Post: 24.88 ± 9.14N; $p = 0.001$, Right lateral bending Pre: 64.41 ± 16.19N, Post: 27.17 ± 9.82N; $p = 0.001$, Left axial rotation Pre: 88.73 ± 11.55N, Post: 33.40 ± 29.04N; $p = 0.001$; Right axial rotation Pre: 88.06 ± 11.37N, Post: 23.92 ± 9.20N; $p = 0.001$) in comparison to pre-surgical scoliosis models. In comparison to the healthy models, pre-surgery model presented considerably increased extension (Healthy: 35.83 ± 4.38N, Pre: 60.24 ± 20.74N; $p = 0.027$) and larger right later bending (Healthy: 46.63 ± 6.66N, Pre: 64.41 ± 16.19N; $p = 0.048$) FJF in the fused level. Furthermore, the post-surgery model presented considerably lower FJF in left lateral bending (Healthy: 46.63 ± 6.66N, Post: 24.88 ± 9.14N; $p = 0.021$), right lateral bending (Healthy: 46.63 ± 6.66N, Post: 27.17 ± 9.82N; $p = 0.\ 028$), left axial rotation (Healthy: 76.11 ± 11.05N, Post: 33.40 ± 29.04N; $p = 0.003$) and right rotation (Healthy: 76.11 ± 11.05N, Post: 23.92 ± 9.20N; $p = 0.\ 001$) in the fused level. Likewise, the post-surgery model presented considerably increased extension (Healthy: 35.83 ± 4.38N, Post: 78.30 ± 37.40N; $p = 0.016$) FJF in the adjacent level in comparison to healthy.

Discussions

The purpose of this study was to investigate the effect of adjacent load transfer before and after fusion surgery of lumbar scoliotic spines using FE models based on in vivo CT scans of scoliotic spines. Spinal fusion did alter the biomechanical function of scoliotic spine. Scoliotic spine presented abnormal and asymmetrical kinetics and kinematics behavior in comparison to the healthy spine.

Minor ROM differences in the pre-surgical and post-surgical scoliosis subjects and between the healthy pre-surgical scoliosis subjects were observed. Our results support Rohlmann et al. (2007) who predicted minor increase in ROM at the adjacent level after spinal fusion surgery. Moreover, a meaningless increase in ROM at the adjacent level was observed by Ambati et al. (2015) when he modeled healthy FE spine model fused with pedicle screws. With the scoliotic spine FE models in this study, small but noticeable increases in ROM at the adjacent level after spinal fusion surgery were predicted at all loading directions. This might suggest that a degenerated adjacent intervertebral disc produced a larger increase ROM after rigid fusion surgery than the healthy adjacent disc did. The increase of ROM at the adjacent level after spinal fusion surgery could be explained by basic engineering concepts. The spinal fusion surgery limited the ROM of the fused level, which reduced the number of actual motion segments in the spine model. As a result of the fusion and the effort of the human body to generate and maintain a motion, compensation will occur at the adjacent segment. When subjected to the identical loads with the pre-surgical model, more deformation in the intervertebral disc at the adjacent level is expected. Despite the nonlinear-elastic material properties of the disc, the spine could be modeled as a spring-damper system, in which vertebrae are connected by springs and dampers in series (Bazrgari et al. 2008). After reducing the number of springs in the series, more deformation will be generated at the remaining springs (intervertebral discs). However, the predicted increase of ROM at the adjacent level caused by the spinal fusion surgery was much lower than conceptual estimations, which predicted ROM at adjacent level in the post-surgical to be at least one time higher than that in pre-surgical model. One major cause for this is the highly nonlinear-elastic material properties of the soft tissues in the intervertebral discs. Another reason is that the spring-damper model of the lumbar spine was not able to fully represent the anatomy in the spine, which has high impact on the biomechanical characterizations of spine (Schmidt et al. 2013).

The average increase percentages of IDP at the adjacent level after the spinal fusion surgery are higher than those in ROM. Surgical fusion predicted slightly increased IDP in the post-surgical models than those predicted in the pre-surgical models at the adjacent level and considerably decreased IDP at the fused level. Minor differences were found between the pre- and post-surgery at the adjacent level in comparison to the healthy. As expected, spinal fusion surgery produces significantly lower IDP at the fused level in comparison to the healthy. The FJF in the scoliotic spine was found to be increased at the adjacent level and considerably decreased FJF at the fused level due to a surgical fusion. Minor differences were found between pre- and post-surgery at the adjacent level in comparison to the healthy. As expected, spinal fusion surgery produces much lower FJF at the fused level in comparison to the healthy.

Our results are consistent with the prediction in previous studies in the literatures (Ambati et al. 2015; Rohlmann et al. 2007) which suggested that spinal fusion surgery has a larger impact on IDP and FJF than on ROM at the adjacent level. However, comparing with the literatures (Hilibrand and Robbins 2004), the increase in IDP and FJF at the adjacent level observed after fusion surgery was not as dramatic. This observation was consistent with the prediction by Rohlmann et al. (2007) using the distracted intervertebral model. The FJF depends on the geometry of the spinal facet joints (Dreischarf et al. 2014). Due to the abnormal spinal curvature in scoliotic spine, the FJF predicted by the pre-surgical scoliosis subjects were much larger than that in the healthy subjects except for the axial rotation loading modes. After the spinal fusion surgery, FJF increased considerably in axial rotations whereas no significant increase was observed at extension and lateral bending loading modes. Rohlmann et al. (2007) also observed the increase of FJF in extension at adjacent level after the fusion surgery, where they explained that this was due to the modified rotational angles after the insertion of an implant.

After the fusion surgery, extension produced the largest ROM in the fused level among the six loading modes whereas flexion and lateral bending generated greater ROM than extension in the pre-surgical scoliosis and healthy subjects. This might suggest that fusion surgery prevent less ROM in extension than that in other loading modes. This suggested that spinal rigid fusion surgery was able to substantially reduce the load applied on the FJF at the fused level, which may relief the pain in facet joints and reduce the risk of further facet joint injuries at the fused level (Davis 2013). At the fused level, the percentage differences in ROM between healthy and pre-surgical subjects were significantly lower than those in IDP and FJF. This suggested that the IDP and FJF increase in pre-surgical scoliosis subjects were not solely caused by the increase in ROM. Other factors such as abnormal spinal anatomy in the scoliotic spines and degenerated intervertebral discs might also contribute to the increase of IDP and FJF in pre-surgical scoliotic spines compared to the healthy spines.

References

Ambati DV, Wright EK Jr, Lehman RA Jr, Kang DG, Wagner SC, Dmitriev AE (2015) Bilateral pedicle screw fixation provides superior biomechanical stability in transforaminal lumbar interbody fusion: a finite element study. Spine J 15(8):1812–1822

Barrey C, Darnis A (2015) Current strategies for the restoration of adequate lordosis during lumbar fusion. World J Orthop 6(1):117–126

Bazrgari B, Shirazi-Adl A, Kasra M (2008) Computation of trunk muscle forces, spinal loads and stability in whole-body vibration. J Sound Vib 318:1334–1347

Bloemker KH, Guess TM, Maletsky L, Dodd K (2012) Computational knee ligament modeling using experimentally determined zero-load lengths. Open Biomed Eng J 6:33–41

Bowden AE, Guerin HL, Villarraga ML, Patwardhan AG, Ochoa JA (2008) Quality of motion considerations in numerical analysis of motion restoring implants of the spine. Clin Biomech (Bristol, Avon) 23(5):536–544

Davis CG (2013) Mechanisms of chronic pain from whiplash injury. J Forensic Leg Med 20 (2):74–85

Dreischarf M, Zander T, Bergmann G, Rohlmann A (2010) A non-optimized follower load path may cause considerable intervertebral rotations. J Biomech 43:2625–2628

Dreischarf M, Rohlmann A, Bergmann G, Zander T (2011) Optimised loads for the simulation of axial rotation in the lumbar spine. J Biomech 43:2625–2628

Dreischarf M, Zander T, Shirazi-Adl A et al (2014) Comparison of eight published static finite element models of the intact lumbar spine: predictive power of models improves when combined together. J Biomech 47(8):1757–1766

Fedorov A, Beichel R, Kalpathy-Cramer J et al (2012) 3D Slicer as an image computing platform for the quantitative imaging network. Magn Reson Imaging 30(9):1323–1341

Hilibrand AS, Robbins M (2004) Adjacent segment degeneration and adjacent segment disease: the consequences of spinal fusion? Spine J 4(6 Suppl):190S–194S

Kotwal S, Pumberger M, Hughes A, Girardi F (2011) Degenerative scoliosis: a review. HSS J 7 (3):257–264

Kotwicki T, Chowanska J, Kinel E, Czaprowski D, Tomaszewski M, Janusz P (2013) Optimal management of idiopathic scoliosis in adolescence. Adolesc Health Med Ther 4:59–73

Little JP, de Visser H, Pearcy MJ, Adam CJ (2008) Are coupled rotations in the lumbar spine largely due to the osseo-ligamentous anatomy? A modeling study. Comput Methods Biomech Biomed Eng 11(2):95–113

Rohlmann A, Burra NK, Zander T, Bergmann G (2007) Comparison of the effects of bilateral posterior dynamic and rigid fixation devices on the loads in the lumbar spine: a finite element analysis. Eur Spine J 16(8):1223–1231

Schmidt H, Galbusera F, Rohlmann A, Zander T, Wilke HJ (2012) Effect of multilevel lumbar disc arthroplasty on spine kinematics and facet joint loads in flexion and extension: a finite element analysis. Eur Spine J 21(Suppl 5):S663

Schmidt H, Galbusera F, Rohlmann A (2013) What have we learned from finite element model studies of lumbar intervertebral discs in the past four decades? J Biomech 46:2342–2355

Wang L, Zhang B, Chen S, Lu X, Li ZY, Guo Q (2016) A validated finite element analysis of facet joint stress in degenerative lumbar scoliosis. World Neurosurg

Xu M, Yang J, Lieberman IH, Haddas R (2016) Lumbar spine finite element model for healthy subjects: development and validation. Comput Methods Biomech Biomed Eng 20:1–15

Zheng J, Yang Y, Lou S, Zhang D, Liao S (2015) Construction and validation of a three-dimensional finite element model of degenerative scoliosis. J Orthop Surg Res 10:189

A Finite Element Platform for Helmet Efficacy Evaluation of Non-Penetrating Projectile Impacts

Rinat Friedman, Ayelet Haimy, Yoram Epstein and Amit Gefen

Abstract The present study introduces a quantitative and methodological tool for assessing the effectiveness with which ballistic helmets protect the brain and optic nerve against non-penetrating projectile impacts. A three-dimensional finite element biomimetic head model with different ballistic helmets was used for this purpose.

Introduction

Traumatic brain injury (TBI) is a damage to the brain due to external mechanical forces, and/or a fast head movements caused by rapid acceleration, deceleration, or rotation. In the last few years, there has been a significant increase in the prevalence of head injuries among soldiers, many of which are due to shrapnel and projectiles impacts. A non-penetrating impact of a bullet/fragment with a combat helmet causes TBI by passage of kinetic energy inwards into the skull, generating fast deformations, stretching, shearing, and eventual collapse of the intracranial soft tissues. Helmet testing methods vary between manufacturers, making it highly difficult to compare helmets designed by different companies (Friedman et al. 2016). There is still no efficient, low-cost, and standardized procedure for assessing the effectiveness of a ballistic helmet.

Our objective was to develop a quantitative methodological tool for a precise and comprehensive assessment of the effectiveness of a ballistic helmet in military injury scenarios in regard to TBI.

R. Friedman · A. Haimy · A. Gefen (✉)
Department of Biomedical Engineering, Tel Aviv University, Tel Aviv, Israel
e-mail: gefen@eng.tau.ac.il

Y. Epstein
Heller Institute of Medical Research, Sheba Medical Center,
Ramat Gan and the Sackler Faculty of Medicine,
Tel Aviv University, Tel Aviv, Israel

A. Gefen and D. Weihs (eds.), *Computer Methods in Biomechanics and Biomedical Engineering*, Lecture Notes in Bioengineering,
DOI 10.1007/978-3-319-59764-5_26

Materials and Methods

The platform consists of several parts: (1) A three-dimensional finite element (FE) biomimetic head model, which includes: skin, fat, skull, jaw, sinus, cerebrospinal fluid (CSF), brain, optic nerves, eyes, spinal cord, vertebrae, and discs was developed. The anatomical model was created based on grayscale transversal cranial images from the Visible Human Project. All tissues were modeled as tided, therefore relative sliding between the skull and the brain was achieved by a very soft and deformable CSF layer. The mechanical properties of the tissues were based on data adopted from the literature. (2) Two ballistic helmet designs were added to the head model with realistic measurements. Both helmets were composed of an external Kevlar®-29 layer and an inner layer of expended polystyrene (EPS) foam. A middle layer for the 3-layered helmet was composed of high-density nylon matrix or EPS. Helmet materials were given bulk (macro) properties, which were adopted from the literature. (3) A 5.56 mm bullet was modeled (standard M-16 ammunition), as being made of 4340 steel. The different components of the computational model are presented in Fig. 1. A collision between the bullet and the two helmets was modeled, and presented in Fig. 1h. The complete 3D model was constructed and meshed with Simpleware ScanIP (Version 6). The dynamic response of the helmet and all the head structures to the impact was simulated. The biomechanical metrics: effective (von Mises) stress, speed, pressure, and acceleration that developed in the intracranial tissues after the impact of the bullet with the helmet were quantitatively evaluated. The FEBio Software Suite (University of Utah, US) was used for assigning material properties, boundary conditions, and speed of impact (PreView ver. 1.18.2), for solving the numerical calculations (FEbio ver 2.3.1) and

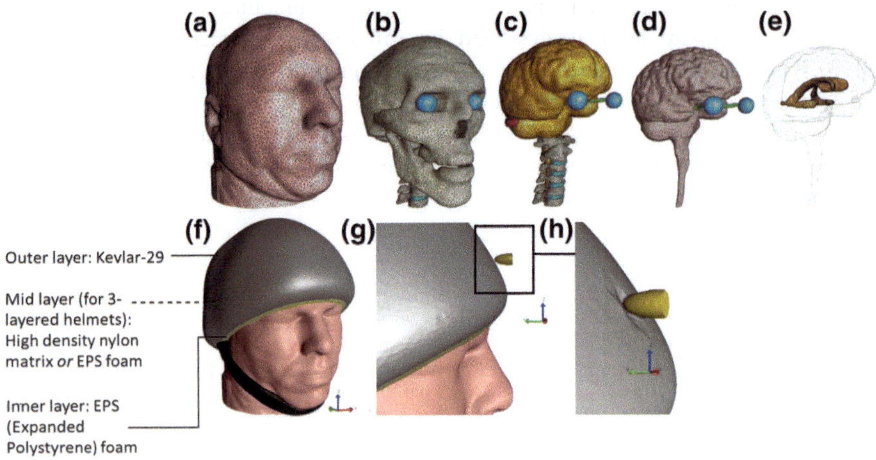

Fig. 1 **a–e** Included tissues in the head model: skin, fat, skull, eyes, optic nerves, cervical spine and vertebrae, disks, bridging veins, CSF, brain, ventricles, **f** Head model with helmet. Helmet materials are presented in the figure, **g** a moment before impact, **h** Bullet collision with the helmet

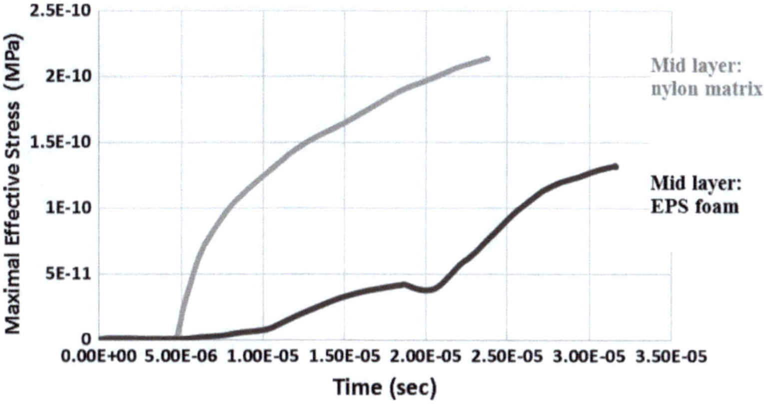

Fig. 2 Maximal effective stress in the brain post impact for two types of helmets—a middle hard nylon layer (*light grey*) and a middle layer of EPS foam (*dark grey*)

for post-processing of the results (PostView ver. 1.9.1), e.g., stress and strain analysis of the cranial tissues after each impact scenario.

Results

Our results indicate that lowering the stiffness of the mid-layer of the helmet can significantly increase the protection capabilities of the helmet Increasing the thickness of the sponge layer of the helmet lowers the stress levels over the surface of the helmet, but does not reduce the resultant in-cranial stress levels. Reducing the Young's modulus (E) of the outer Kavlar-29 layer by half, improves its protection capabilities, while decreasing E by 80% or the shear modulus G of this layer by 85% does not promote protection of the brain. In addition, the two types of the 3-layered helmets were compared to each other—one with a stiff nylon matrix middle layer and another with a EPS foam middle layer. Figure 2 shows this comparison in terms of maximal effective stress in the brain following the impact with the helmet. Figure 3 provides effective stress color maps of the brain, CSF, and eyes with optic nerve, when comparing the results for the two three-layered helmets.

Discussion

Using the current settings for helmet structure and mechanical properties, we proved that a higher elastic modulus for the helmet shell does not necessarily provide a greater protection. The same is not true for the shear modulus (G),

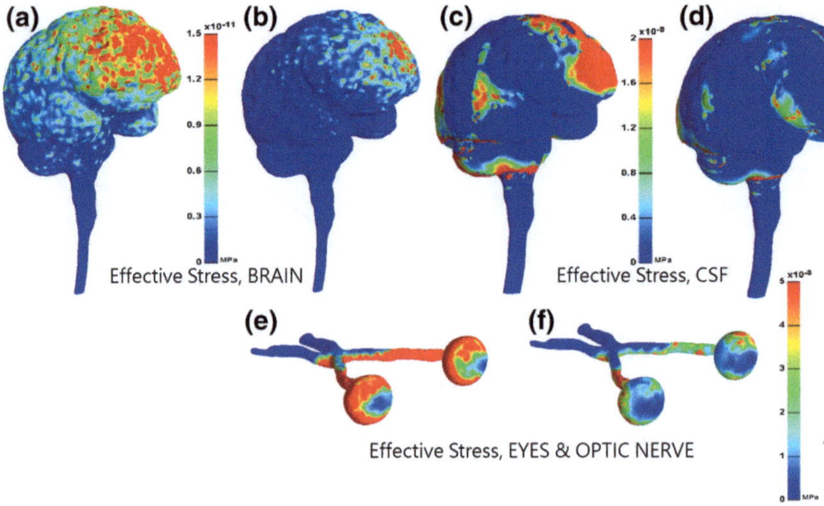

Fig. 3 Effective stress in various cranial tissues post impact, for the helmets described in Fig. 1, arranged in pairs of Nylon—EPS, and tested under the same scenario of impact. **a**, **b** Effective stress in the brain with the Nylon versus EPS helmet. **c**, **d** Effective stress in the CSF with the Nylon versus EPS helmet. **e**, **f** Effective stress in the optic nerve with the Nylon versus EPS helmet

as lowering its level does not contribute to the reduction of effective stress levels in the brain. In this work, we have demonstrated some of the capabilities provided by our platform for the comparative, quantitative, and cost-effective evaluation of various helmet designs, by allowing to easily change various parameters in the helmet and then quantitatively compare the effects of these changes on the inter-cranial forces.

Acknowledgements This work was funded by IDF's Research and Development Agency—Mafaat and the Medical Corps.

Reference

Friedman R, Epstein Y, Gefen A (2016) Traumatic brain injury in the military: biomechanics and finite element modelling, in studies in mechanobiology. In: Tissue engineering and biomaterials. Springer, Berlin, pp 1–25. doi:10.1007/8415_2016_189

Tissue Loads Applied by a Novel Medical Device for Closing Large Wounds

Rona Katzengold, Moris Topaz and Amit Gefen

Abstract Closure of large soft-tissue defects following surgery or trauma constitutes substantial but common reconstructive challenges. Closing the wounds with sutures is a common solution yet involving high-tension closure. The alternative methods of closure such as skin grafting are often associated with relatively more complex surgical procedures, significant morbidity, and extended hospitalization and recovery periods. Here, we evaluate the efficacy of a tension relief system (TRS) device and compare it to surgical sutures. We employed finite element modeling and simulated three cases of (real) large wounds which were treated with TRS in reality, each located in a different organ and has different dimensions. Closure of the wounds induced peak-effective stresses on the skin that were at least an order of magnitude greater (and sometimes nearly 2 orders of magnitude greater) than when tension sutures were used with respect to the corresponding TRS data. For the tension suture simulations, the tensile stress was in the range of 415–648MPa and in the TRS simulations, it was 16–30MPa. Such large localized tissue distortions may obstruct the vasculature surrounding the wound or at the sutured skin itself, which will cause ischemia and necrosis of the skin within several hours following surgery. In addition, the substantial reduction of loads on and within the skin during large wound closure by the TRS allows surgeons to optimally employ the viscoelastic properties of the skin for primary wound closure.

R. Katzengold (✉) · A. Gefen
Department of Biomedical Engineering,
Faculty of Engineering, Tel Aviv University, Tel Aviv, Israel
e-mail: ronakatzengold@gmail.com

M. Topaz
Plastic Surgery Unit, Hillel Yaffe Medical Center, Hadera, Israel

A. Gefen and D. Weihs (eds.), *Computer Methods in Biomechanics and Biomedical Engineering*, Lecture Notes in Bioengineering,
DOI 10.1007/978-3-319-59764-5_27

Introduction

Closure of large soft tissue defects following surgery or trauma constitutes substantial but common reconstructive challenges (Topaz et al. 2014). Closing wounds with sutures has been the gold standard approach since the early days of surgery, yet it may be impeded and limited, particularly with regard to large wounds (Al-Mubarak and Al-Haddab 2013). In such cases, an attempt to use conventional suturing will result in high-tension closure, and therefore, alternative external skin stretching systems (TRS) were developed. These types of devices were meant to reduce local mechanical loads in skin and underlying tissues, and allow taking advantage of the viscoelastic properties of the skin (Topaz et al. 2012).

Studies have shown the clinical advantages of skin stretching systems, however, quantitative bioengineering models, demonstrating closure of large wounds, are lacking.

Here, we present a finite element (FE) modeling of the TRS and its biomechanical efficacy in three (real) wound cases, compared with the alternative of a conventional surgical suturing closure technique, which has not been used in these reported cases.

Methods

To make the modeling and simulations more realistic and clinically relevant, we modeled the TopClosure® TRS in its application as an external tissue stretching system in three specific large wounds. The TopClosure® comprises of two polymer attachment plates (APs) that are fixed to the skin and are pulled together manually by an interconnecting flexible approximation strap (AS) (Fig. 1a). Each of the three wounds, in reality, was located in a different body region, in three different patients. The geometrical models of the tissue defects were built per each wound case using the Scan-IP module of Simpleware® and then the APs of the TRS were added according to the number of APs that were used in practice, to close the real specific wound. For the analyses of wound closure using tension sutures, 0.6 mm-thick half-rings were simulated, matching the diameter of the tension sutures that would have been used in these cases, in both sides of the wound (Fig. 1b). For the purpose of a sensitivity analysis, we created three additional wound model configurations, each containing additional sutures. These last three models were built in order to measure the effects of an optional solution in each of these wound situations, which is to close the wounds with more sutures. Both the APs and the tension sutures were displaced in the horizontal direction toward closing the wound (Fig. 1).

The wound closure models were analyzed for Lagrange stresses. Outcome measures included: (1) Distributions of effective stresses in the tissue VOI. (2) Peak effective stresses in the tissues. (3) Tensile stress in the tissues. These outcome

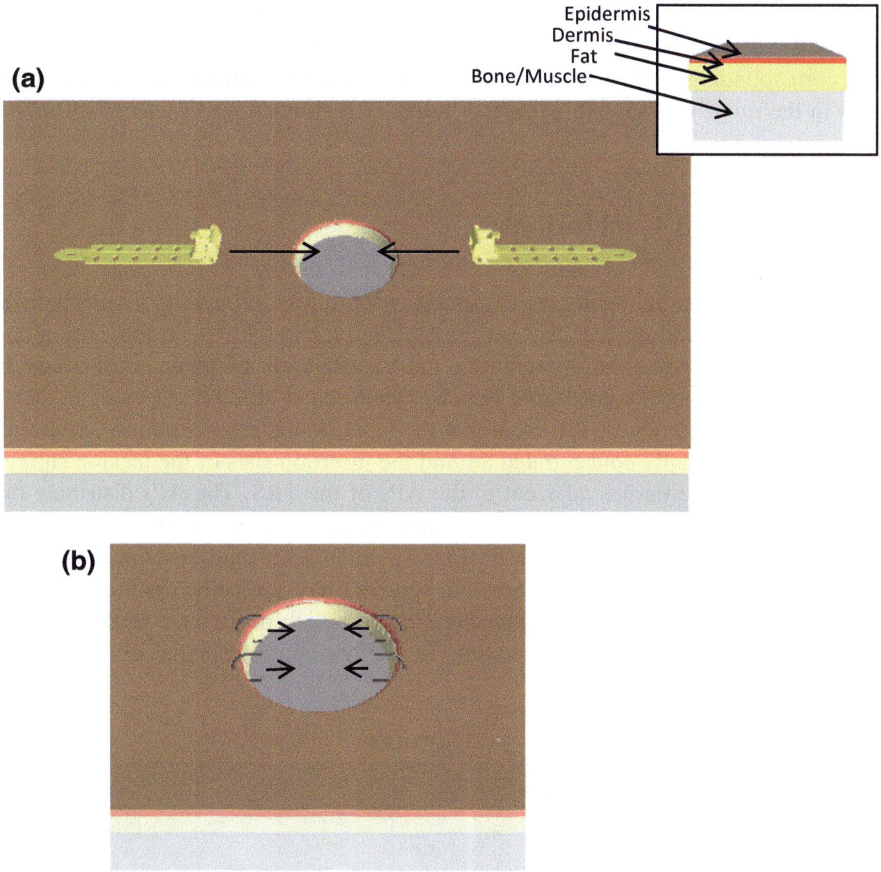

Fig. 1 The configurations of the finite element models, with boundary and loading conditions: **a** Wound closure using the TopClosure® tension relief system (IVT Medical Inc., Ra'anana, Israel). The layered structure of the tissues represented in the model is depicted in *the top right panel*. **b** Wound closure using tension

measures were compared between the TRS closure option (which was taken in reality) and the suture closure option in each wound case.

Results

Closure of all the wounds by both the TRS system and tension sutures induced localized stresses in the skin and deeper tissues. These stresses reached maximum values around the APs for the TRS closures as well as around the insertion sites of the sutures, mostly in the side opposite to the wound due to the pulling activity of the two closure techniques (Fig. 2). Closure of the wounds induced peak effective stresses on the skin that were at least an order of magnitude greater (and sometimes nearly 2 orders of magnitude greater) when tension sutures was used with respect to the corresponding TRS data. For the tension suture simulations, the tensile stress was in the range of 415–648 MPa and in the TRS simulations, it was 16–30 MPa.

Discussion and Conclusions

Computational FE modeling is extremely useful for comparing biomechanical efficacies of medical devices and surgical techniques in the wound prevention and treatment arena, particularly regarding the distributions of mechanical loads in tissues, which cannot be measured directly (Halloran et al. 2005; Levy et al. 2014, 2015). From Fig. 2, Query it is clear that the local skin stresses are much greater in magnitude and focally concentrated around the insertion sites of the tension sutures with respect to the tissues adjacent to the APs of the TRS. The APs distribute the deformations more uniformly, and hence, magnitudes of tissue stresses around them are 1–2 orders of magnitude lower (Fig. 2). Such large and sustained localized tissue deformations caused by tension sutures may critically distort cells and obstruct the vasculature surrounding the wound, which may cause necrosis of the skin within several hours post-operation. In addition, the substantial load reduction

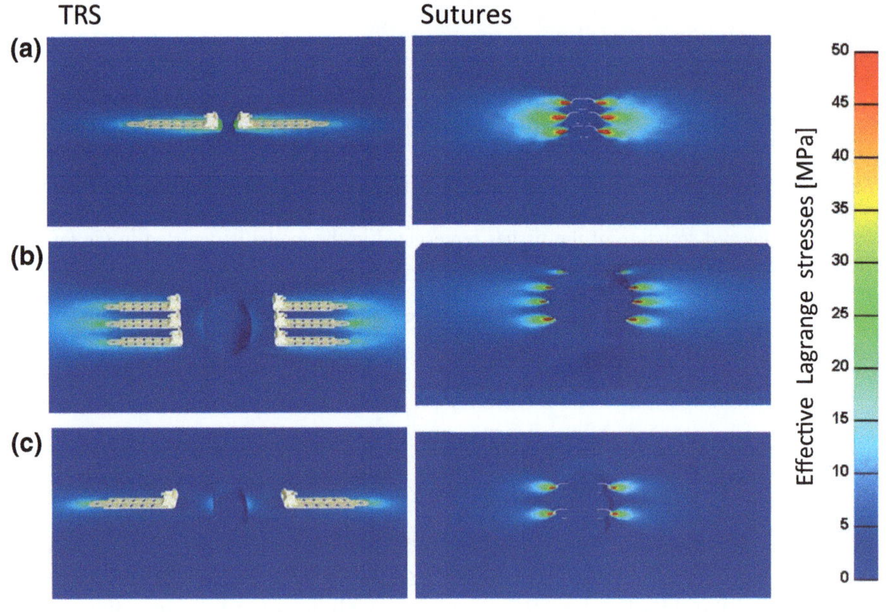

Fig. 2 Effective Lagrange stress distributions on the skin when the TopClosure® tension relief system (TRS) is being used (left frames coulomb) with respect to use of tension sutures in the same wounds (right frames coulomb) for: **a** the scalp wound, **b** the back wound and **c** the leg wound

by the TRS during large wound closure allows surgeons to optimally employ the viscoelastic properties and biological adaptation of the skin.

References

Al-Mubarak L, Al-Haddab M (2013) Cutaneous wound closure materials: an overview and update. J Cutan Aesthet Surg 6(4):178–188

Halloran JP, Petrella AJ, Rullkoetter PJ (2005) Explicit finite element modeling of total knee replacement mechanics. J Biomech 38(2):323–331

Levy A, Kopplin K, Gefen A (2014) Computer simulations of efficacy of air-cell-based cushions in protecting against reoccurrence of pressure ulcers. J Rehabil Res Dev 51(8):1297–1319

Levy A, Frank MB, Gefen A (2015) The biomechanical efficacy of dressings in preventing heel ulcers. J Tissue Viability 24(1):1–11

Topaz M, Carmel NN, Silberman A, Li MS, Li YZ (2012) The TopClosure 3S system, for skin stretching and a secure wound closure. Eur J Plast Surg 35:533–543

Topaz M, Carmel NN, Topaz G, Zilinsky I (2014) A substitute for skin grafts, flaps, or internal tissue expanders in scalp defects following tumor ablative surgery. J Drugs Dermatol 13:48–55

Biomechanics of Implant Failure After PSO: Influence of the Hardware Configuration Through a Finite Element Analysis

Tomaso Villa, Claudia Ottardi, Luigi La Barbera, Andrea Luca and Fabio Galbusera

Abstract Pedicle Subtraction Osteotomy (PSO) is a technique for restoring the correct sagittal balance that achieves corrections up to 35°. The procedure still shows a high rate of complications (10–25%, usually rod breakage at the osteotomy level): in order to decrease this rate a computational comparative analysis investigating the effects of a set of variables on the instrumentation has been set up. In particular the number of rods, the material, the diameter, and the presence of an anterior support have been considered in an FE model of L1-S1 simulating a PSO at either L3 or L4. Standing was simulated using a follower load of 500 N and pure moments of ± 7.5 Nm in flexion, extension, lateral bending, and axial rotation evaluating the load sharing between the anterior part of the spine and the rods and the stress in the rods. Analyzing our results we can suggest the use of a multiple rod configuration, associated with the use of an anterior support in order to reduce the stresses on the instrumentation.

Introduction

Pedicle Subtraction Osteotomy (PSO) is a common and effective surgical technique used in case of severe sagittal imbalance. A bony wedge and the posterior processes are removed from the vertebra at the apex of the curve (usually on L3 or L4), obtaining a correction of the curvature up to 30°–35° (Bridwell et al. 2003). This surgical technique is associated to a long spinal instrumentation, at least two levels

T. Villa (✉) · C. Ottardi · L. La Barbera
Laboratory of Biological Structure Mechanics, Department of Chemistry, Materials and Chemical Engineering “G. Natta”,
Politecnico di Milano, Milan, Italy
e-mail: tomaso.villa@polimi.it

T. Villa · L. La Barbera · F. Galbusera
IRCCS Istituto Ortopedico Galeazzi, Milan, Italy

A. Luca
Department of Spine Surgery III, IRCCS Istituto Ortopedico Galeazzi, Milan, Italy

A. Gefen and D. Weihs (eds.), *Computer Methods in Biomechanics and Biomedical Engineering*, Lecture Notes in Bioengineering,
DOI 10.1007/978-3-319-59764-5_28

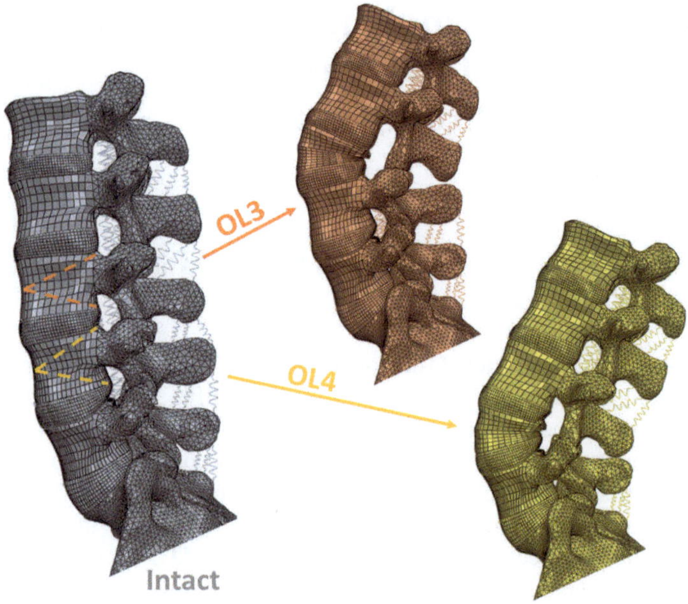

Fig. 1 Computational model of the intact spine (*left*) and osteotomy models (*right*)

above and below the osteotomy site. Due to its complexity, a high risk of instrumentation failure is reported in the literature (Luca et al. 2014).

Since in literature only few studies investigate the biomechanics of PSO with computational of experimental models (Charosky et al. 2014; Tang et al. 2013; Hallager et al. 2016; Deviren et al. 2012), the goal of this work is to fill this gap through a computational comparative analysis that investigates the effects of a set of variables in the instrumentation in order to decrease its rate of failure.

Materials and Methods

A nonlinear finite element model of the lumbar spine ranging from L1 to S1 was created and calibrated (Ottardi et al. 2016). Subsequently, a PSO was simulated on two levels: L3 and L4 (Fig. 1) (Ottardi et al. 2016). All the models were afterwards instrumented using polyaxial pedicle screws and intersomatic cages (Fig. 2), designed using the CAD software PTC Creo Parametric 2.0. A sensitivity analysis was executed on the mesh of each device and the analysis were performed using Abaqus 6.12-3 (Luca et al. 2016). The condition of standing was simulated using a follower load of 500 N and pure moments of ± 7.5 Nm in flexion, extension, lateral bending and axial rotation evaluating the load sharing between the anterior part of the spine and the rods and the stress in the rods.

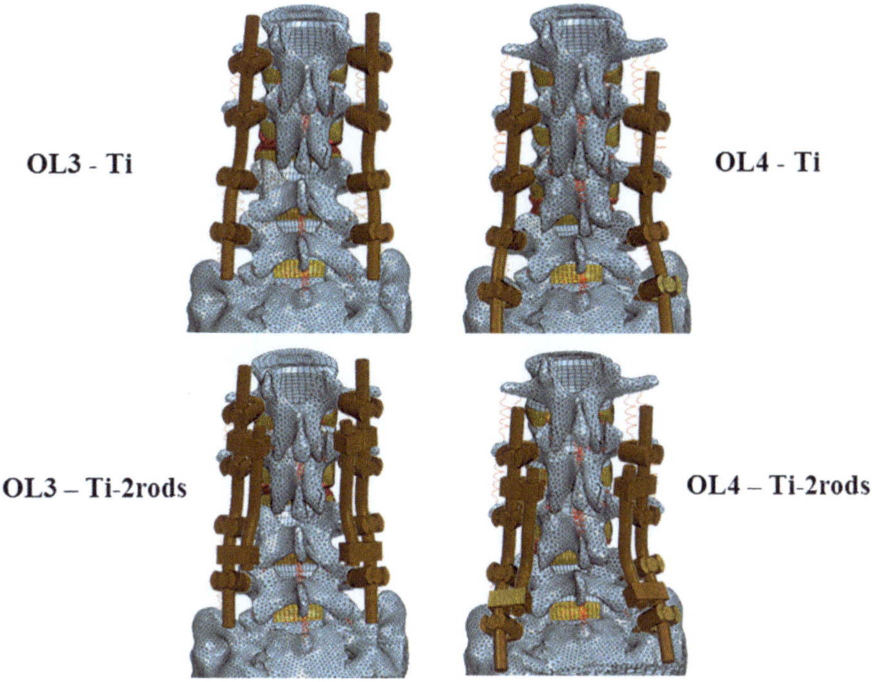

Fig. 2 Osteotomy models after the stabilization with spinal fixators

The influence of the following parameters on the stress arising in the rods has been investigated: (i) number of rods (one or two, Fig. 2), (ii) material of the rods (Ti or CoCr alloy), (iii) diameter of the rods (5 or 6 mm), (iv) anterior support (TLIF or XLIF and their position, Fig. 3). The configuration with one 6 mm rod in Ti-alloy without any anterior support was assumed as a reference.

Results

The main stress concentration on the rods is at the osteotomy level. In all configurations, the maximum stress value was well below the yielding point typical for Ti- and CoCr-alloys (Fig. 4).

A stiffer material increases the load (7%) and the stress on the rods (about 35%). A smaller diameter induces an 8% decrease of the load on the rod but increases stress at the osteotomy level up to 20%. The configuration with two rods increases the load on the instrumentation (10%) with a decrease of stress on the principal rods up to 50%. Therefore, Co–Cr double parallel rods produce the higher stress reduction on the implants near the treated level. A very good qualitative agreement was found for models reproducing a PSO at L3 and L4 levels, therefore, subsequent analysis was continued only on OL3 model.

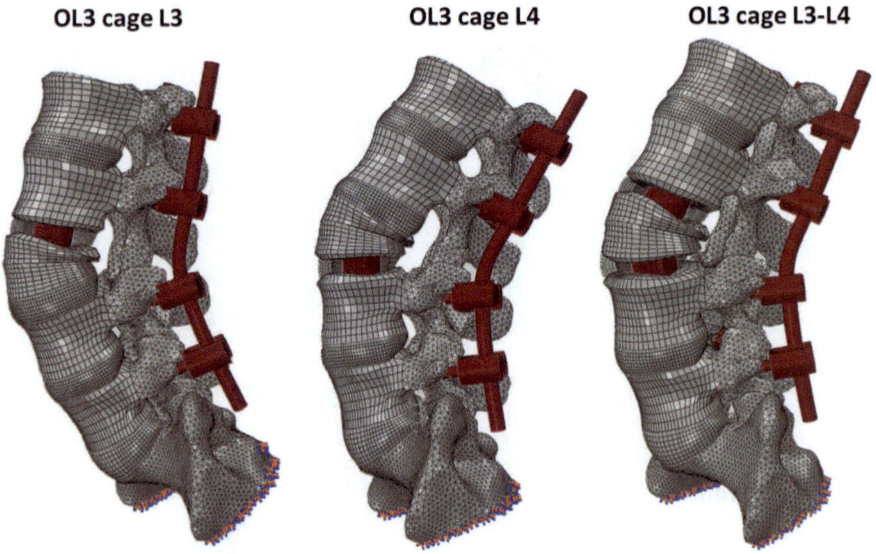

Fig. 3 Osteotomy models after the insertion of a cage above (*left*), below (*middle*) or above and below the osteotomy (*right*)

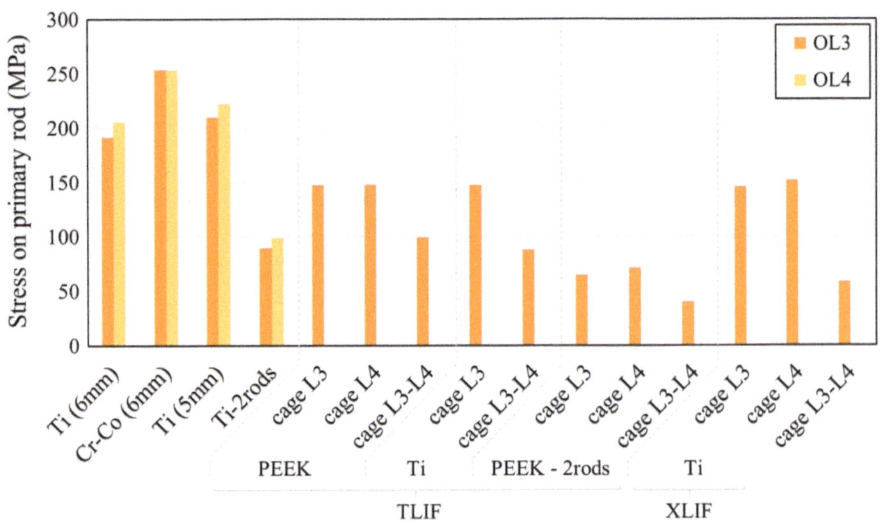

Fig. 4 Stress on the primary rods

Moreover, the use of the anterior support is important to preserve the rods: when two cages are used, an important reduction of the stresses is noticed (up to 50% with respect to the instrumented configuration). Marginal differences were found

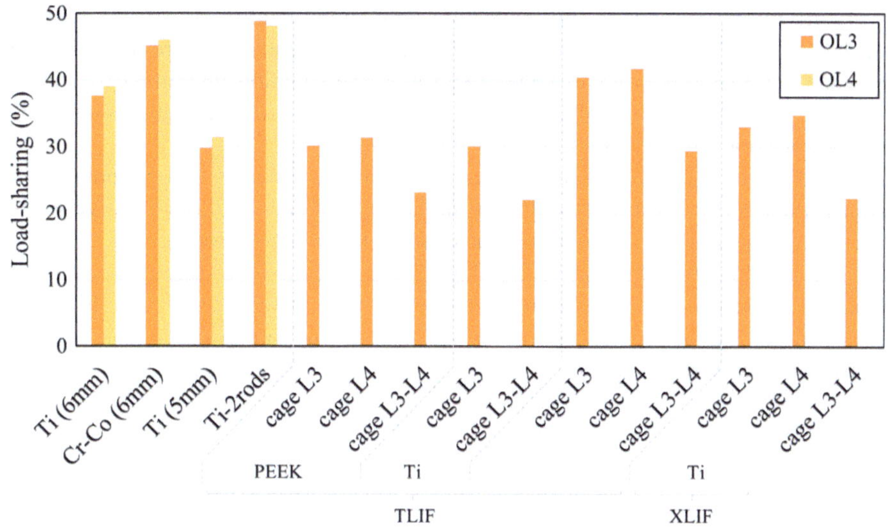

Fig. 5 Load shared by the spinal rod

comparing the position of the cage above or below the osteotomy and different materials (PEEK and Ti6Al4V). Also comparing two cage designs, TLIF versus XLIF, only small stress reductions on the rods (about 10%) were calculated using a larger XLIF cage instead of a TLIF (Fig. 5).

Discussion

In a previous study, the effect of the PSO performed on different levels (L3 and L4) was analyzed (Ottardi et al. 2016). The results showed that PSO induces high destabilization, especially in the lower levels. Therefore, also the spinal fixators when are used in combination with a PSO are highly stressed. The goal of this work is to investigate different configuration of instrumentations in terms of materials, number of rods, diameter of the rods and use of cages, in order to identify a solution able to reduce the stresses at the osteotomy level and prevent the hardware failure.

Cr–Co rods are known to be more resistant to fatigue and with less notch sensitivity than other materials like Ti6Al4V (Smith et al. 2014). Those characteristics may play a crucial role during PSO surgeries, where rods are bended up to 60°. However, our results showed that the stiffer material induces a load and stress increase on the rods, which may influence the fatigue properties. Using smaller rods in terms of diameter causes a load reduction due to the decreased stiffness of the system. However, an increase of the stress was found, probably due to the change of the resistance section.

Using intersomatic cages in association with spinal fixators, an overall increase of the stiffness of the system was calculated. Consequently, the stress on the rods resulted significantly lower in comparison with the situation instrumented without anterior support. Remarkably, different cage designs and materials caused only minor variation on the load and stress distribution.

The limitations regarded the contact between the different components of the spinal fixators (tie constraint) and the residual stresses causes intra-operatively with the bending procedure of the rods that was not taken into account.

As a conclusion, even if several biological factors cannot be simulated with finite element models, analyzing our results we can suggest the use of a multiple rod configuration, associated to the use of an anterior support in order to reduce the stresses on the instrumentation.

References

Bridwell KH, Lewis SJ, Lenke LG (2003) Pedicle subtraction osteotomy for the treatment of fixed sagittal imbalance. J Bone Jt Surg Am 85:454–463

Charosky S, Moreno P, Maxy P (2014) Instability and instrumentation failures after a PSO: a finite element analysis. Eur Spine J 23(11):2340–2349

Deviren V, Tang J-A, Scheer J-K, Buckley J-M, Pekmezci M, McClellan R-T, Ames C-P (2012) Construct rigidity after fatigue loading in pedicle subtraction osteotomy with or without adjacent interbody structural 6 cages. Glob Spine J 2(4):213–220

Hallager DW, Gehrchen M, Dahl B, Harris JA, Gudipally M, Jenkins S, Wu AM, Bucklen BS (2016) Use of supplemental short pre-contoured accessory rods and cobalt chrome alloy posterior rods reduces primary rod strain and range of motion across the pedicle subtraction osteotomy level: an in vitro biomechanical study. Spine 41(7):E388–E395

Luca A, Lovi A, Galbusera F, Brayda-Bruno M (2014) Revision surgery after PSO failure with rod breakage: a comparison of different techniques. Eur Spine J 23(Suppl 6):610–615

Luca A, Ottardi C, Sasso M, Prosdocimo L, La Barbera L, Brayda-Bruno M, Galbusera F, Villa T (2016) Instrumentation failure following pedicle subtraction osteotomy: the role of rod material, diameter, and multi-rod constructs. Eur Spine J. doi:10.1007/s00586-016-4859-8

Ottardi C, Galbusera F, Luca A, Prosdocimo L, Sasso M, Brayda-Bruno M, Villa T (2016) Finite element analysis of the lumbar destabilization following pedicle subtraction osteotomy. Med Eng Phys 38(5):506–509

Smith JS, Shaffrey E, Klineberg E, Shaffrey CI, Lafage V, Schwab FJ, Protopsaltis T, Scheer JK, Mundis GM, Fu K-MG, Gupta MC, Hostin R, Deviren V, Kebaish K, Hart R, Burton DC, Line B, Bess S, Ames CP (2014) Prospective multicenter assessment of risk factors for rod fracture following surgery for adult spinal deformity. J Neurosurg Spine 21(6):994–1003

Tang JA, Leasure JM, Smith JS, Buckley JM, Kondrashov D, Ames CP (2013) Effect of severity of rod contour on posterior rod failure in the setting of lumbar pedicle subtraction osteotomy (PSO): a biomechanical study. Neurosurgery 72(2):276–282

Statistical Shape Modelling to Analyse the Talus in Paediatric Clubfoot

Yixuan Feng, Aaron Bishop, Daniel Farley, Joseph Mitchell, Kenneth Noonan, Xiaoping Qian and Heidi-Lynn Ploeg

Abstract Serial Ponseti casting occasionally fails to completely correct idiopathic clubfoot deformities and recurrence of deformity occurs. As these bones are largely unossified, statistical shape modelling (SSM) from MRI considers the entire set of shape features simultaneously and may provide a more accurate assessment of differences in bone morphology between two groups. The purpose of this study was to compare the shape of the talus in normal and clubfeet using direct geometric measurements and SSM. Both methods found significant differences between the control and clubfoot groups.

Introduction

Clubfoot, also called congenital talipes equinovarus, is one of the most common congenital deformities of the foot (Moritani et al. 2000). The primary abnormality of clubfoot is believed to be a deformity of the talus with displacement of the calcaneus and navicular bone in a medial and plantar direction (Moritani et al. 2000). The evaluation of clubfoot treatment results is mainly dependent on clinical and functional parameters; however, radiological studies are often required in assessment. Magnetic resonance imaging (MRI) is of value to reveal in vivo anatomical relationships in clubfoot deformity without requiring paediatric X-ray

Y. Feng (✉) · D. Farley · X. Qian · H.-L. Ploeg
Department of Mechanical Engineering, College of Engineering,
University of Wisconsin-Madison, Madison, WI, USA
e-mail: yfeng57@wisc.edu

A. Bishop · H.-L. Ploeg
Department of Biomedical Engineering, College of Engineering,
University of Wisconsin-Madison, Madison, WI, USA

J. Mitchell · K. Noonan · H.-L. Ploeg
Department of Orthopedics and Rehabilitation,
School of Medicine and Public Health, University of Wisconsin-Madison,
Madison, WI, USA

A. Gefen and D. Weihs (eds.), *Computer Methods in Biomechanics and Biomedical Engineering*, Lecture Notes in Bioengineering,
DOI 10.1007/978-3-319-59764-5_29

exposure (Pekindil et al. 2001) and may support a better understanding of the clubfoot. Direct geometric measurements have been taken of MRI data in the picture archiving and communication system (PACS) or after construction of a three-dimensional (3D) model; and, a *t* test (Itohara et al. 2005a) or *u* test (Saito et al. 2004) has been performed to statistically analyse the difference of each measurement between normal and clubfoot bone morphology (Moritani et al. 2000; Pekindil et al. 2001; Itohara et al. 2005a, b; Saito et al. 2004; Kamegaya et al. 2001). In previous studies, direct measurements from MRI images were insufficient to accurately assess hindfoot morphology in this population. One limitation was that the axes and planes used to define measurements may have had large variances because they were drawn manually. Large variance and small sample sizes, 7–36, limited the statistical power of these previous studies. However, the main limitation of the direct measurements is that they do not account for the bone's 3D surface morphology (Pekindil et al. 2001). To address some of these issues, statistical shape modelling (SSM) can be used to analyse hindfoot morphology.

Statistical shape modelling (SSM) considers the entire set of shape features simultaneously and may provide a more complete assessment of differences in bone morphology between two groups (Golland et al. 2005). SSM uses shape registration and principal component analysis (PCA) to quantify differences and visually represent each mode that explains the partial differences of the shape between two groups. The relative contributions for all modes can be quantified. And the standardized principal component (Std PC) scores can be calculated for every shape in each mode. SSM is widely used in human-centre product design (Allen et al. 2003; Baek and Lee 2012; Bolkart and Wuhrer 2015; Wuhrer and Shu 2013) and is also applied in medical image analysis on human anatomy, like brain (Gerig et al. 2001; Patenaude et al. 2011; Styner et al. 2006), breast (Rueckert et al. 1999) or knee (Smoger et al. 2015), from MRI data. To-date SSM has not been applied to the clubfoot.

Since the primary abnormality of clubfoot is believed to be a deformity of the talus (Moritani et al. 2000), analysis of the talus can best reflect the shape differences between clubfoot and normal morphology. Therefore, for this study, in addition to direct geometric measurements on 3D models, SSM was conducted to analyse the talus in paediatric clubfoot in patients with residual deformity in comparison to a control group of normal, age-matched ankle bones. The purpose of this study was to develop SSM methods for the analysis of clubfoot morphology. The ability of SSM to differentiate between patient groups was compared with direct geometric measurements.

Materials and Methods

Clinical MRI data was obtained through standard care of 18 children. Talar bones with residual deformity (2.04–6.07 years, average age 4.76 years) were compared with age-matched, control ankles (2.04–6.92 years, average age 4.80 years).

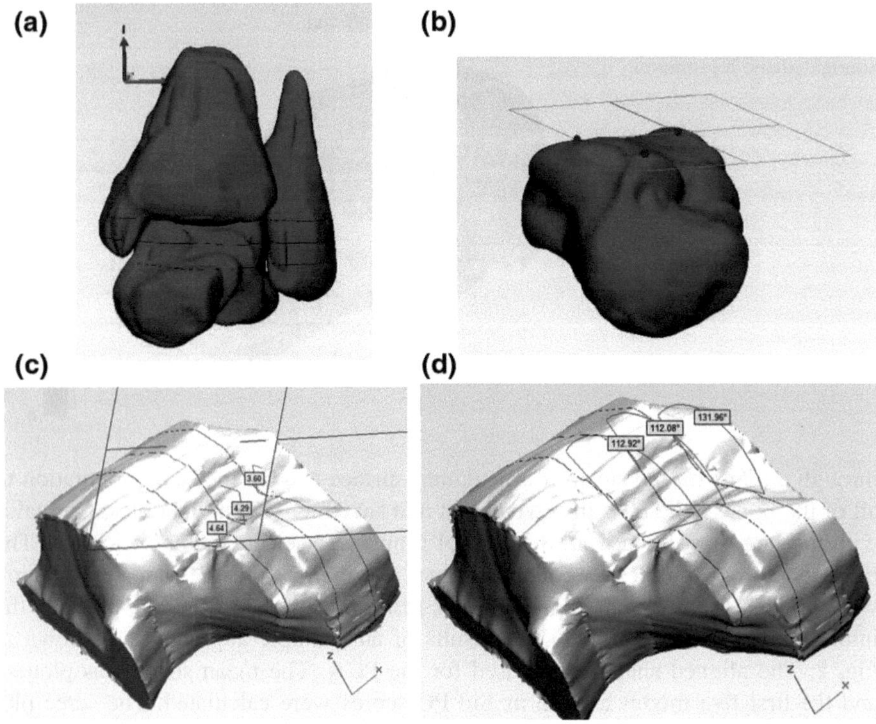

Fig. 1 Summary of measurements on 3D models: **a** full ankle merged; **b** superior surface; **c** neck depth; **d** neck angle

Twenty-one models were involved in the direct measurements method, including 14 clubfoot and 7 control models. SSM was conducted on 26 models, including 16 in the clubfoot group and 10 in the control group. Twenty-three measurements were obtained from each 3D model generated via reverse engineering (Mimics v17, Materialise, Leuven, Belgium). The reverse engineering process of MRI sectioning in Mimics was performed by clinical collaborators to ensure accurate identification of anatomical structures. Each ankle was segmented three times, once with each anatomical plane in high resolution. The three individually segmented models were then combined using Geomagic Studio (3D Systems, Rock Hill, South Carolina) to produce a single model with high resolution in all planes. The combined models where then analysed in 3-Matic (version 10, Materialise, Leuven, Belgium). Seventeen measurements were taken on talar models including seven linear measurements, three angle measurements, three radial measurements and one volume measurement. The remaining nine measurements were of the mortise. Examples of the model measurements are shown in Fig. 1.

SSM, using nonrigid shape registration and principal component analysis with a custom-made code (Matlab R2015a, Mathworks, Natick, MA), was conducted on 26 talar models. All the left ankle talar STL models were mirrored in 3-Matic

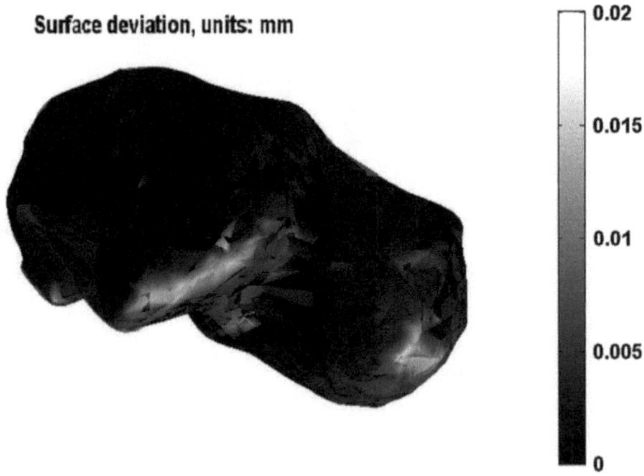

Fig. 2 Lateral view of template and talus after nonrigid shape registration

through a plane on the superior articulating surface to give the same orientation to all of the bones. The STL files were read and saved as Matlab data files. One talus with the largest volume from the control group was selected as the template. The rigid shape registration was conducted first, and then a nonrigid shape registration was applied. The resultant aligned shapes from the template were compared to the initial models from STL files. The results of an example alignment are shown in Fig. 2. The aligned shapes were used for the PCA. The mean shape was plotted, and the first five modes as well as Std PC scores were calculated. The scree plot showing the contribution of each mode was created. The variation inside each mode from −3 standard deviations to +3 standard deviations was found and plotted. The PCA was performed for the clubfoot group and control group. The modes from these two different groups were compared. The Std PC scores from clubfoot and control groups were compared using Student *t* test.

Results

Twenty-one models were involved in the direct measurements method, 7 normal and 14 clubfoot models. The mean value and standard deviation of each parameter were calculated for both groups. The *t* test assuming unequal variances was conducted among two groups and the *p* value based on two-tail test was reported in Table 1.

From Table 1, the mean Superior Width in clubfeet was 4.43 mm (23%) less than the control group ($p < 0.05$). The average Neck Angle 2 was greater in clubfeet (147°) than in controls (132°, $p < 0.05$). The Talar Neck Depth 2 was 48% less in the clubfeet (1.86 mm) in comparison to the control group (3.88 mm, $p < 0.01$). Among 17 measurements, 12 of them were larger among the normal group. And the standard deviation was larger in clubfoot for 11 out of 17 measurements. Also, nine of these measurements indicated significant differences between the control and clubfoot

Table 1 Mean value, standard deviation and p value of direct measurements, asterisks indicate $p < 0.05$

Direct measurement	Mean value		Standard deviation		p value
Group	Normal	Clubfoot	Normal	Clubfoot	$(T \leq t)$ two-tail
Superior width	19.01	14.58	4.57	2.73	0.04556*
Halfway to Medial Malleolus	27.07	22.79	1.22	3.98	0.001809*
Medial Malleolus	29.37	25.98	1.06	4.44	0.01547*
Superior width 2 (widest)	18.96	15.37	4.31	3.17	0.08232
Halfway width 2 (widest)	26.09	23.66	1.35	4.15	0.06208
Medial width 2 (widest)	27.73	26.66	2.01	5.02	0.4964
Length	43.40	42.49	3.91	6.00	0.6807
Talar neck 1	3.19	2.38	1.23	1.15	0.1717
Talar neck 2	3.88	1.86	1.28	0.67	0.004368*
Talar neck 3	3.93	2.14	0.83	0.91	0.0005994*
Neck angle 1	154.14	146.60	14.81	13.96	0.2866
Neck angle 2	132.09	146.75	13.55	11.35	0.03351*
Neck angle 3	126.49	147.78	12.92	14.09	0.004287*
Volume	29,338	21,669	7797	8908	0.06247
Arc talus 1	17.25	24.45	6.43	8.32	0.04521*
Arc talus 2	16.95	20.59	3.15	4.04	0.03912*
Arc talus 3	17.84	20.42	2.01	7.38	0.2399

The unit of all linear measurements is mm. The unit of all volume measurements is mm^3. The unit of all angle measurements is degree

groups ($p < 0.05$). This indicated that direct measurements found significant differences between the control and clubfoot groups.

SSM, including nonrigid shape registration and principal component analysis, was conducted on 26 models, 16 in the clubfoot group and 10 in the control group. The "Talus #3" bone was chosen as the template shape since it had the largest volume. The number of vertices of the template was reduced to 4725 in order to run the analysis on a desktop computer (16 GB RAM, Intel Xeon CPU E5405 2.00 GHz, 2 processors). All the other target shapes were also reduced to contain similar numbers of vertices.

The aligned template and target shape were plotted after nonrigid shape registration (Fig. 2). The light-coloured shape represents the target shape; and, the solid dark-coloured shape represents the aligned template shape. The mean value of surface deviation was 0.0022 mm and the maximum surface deviation was about 0.016 mm, which were much better than the results after rigid shape registration.

The PCA was conducted on the control group and the clubfoot group. The volumes of all models were normalized to 1 so that the influence of size was excluded. The first five modes explained about 63% of the variation among 26 models. The first 15 principal components could explain about 90% of the variation and 20 principal components could explain about 95% of the variation. The lateral

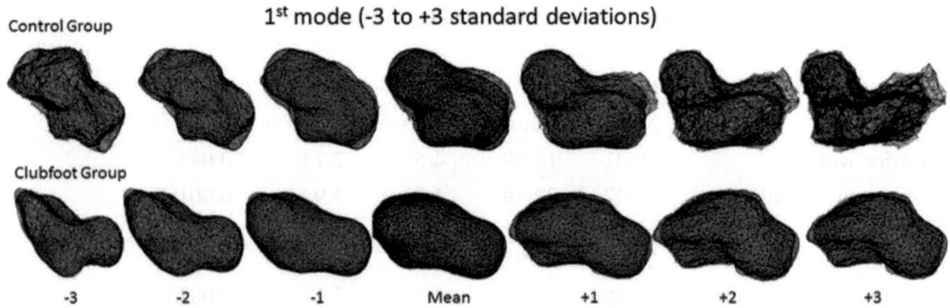

Fig. 3 First principal component mode

Table 2 Mean values (± standard deviation) of Std PC scores

Group	1st Std PC scores	2nd Std PC scores	3rd Std PC scores	4th Std PC scores	5th Std PC scores
Control	−0.47 ± 0.98*	0.09 ± 0.29	−0.21 ± 0.86	0.64 ± 0.68**	0.35 ± 0.91
Clubfoot	0.29 ± 0.92*	−0.05 ± 1.27	0.13 ± 1.09	−0.40 ± 0.97**	−0.22 ± 1.02

Asterisks indicate the mean differences between two groups were significant

Fig. 4 First Std PC scores versus fourth Std PC scores

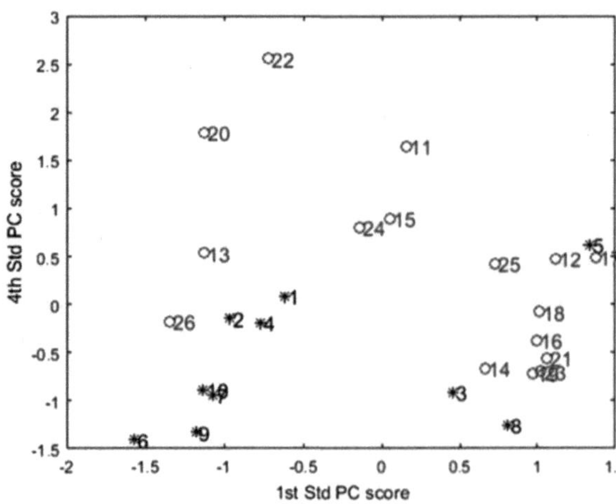

view of the talus in first mode from −3 deviations to +3 deviations is shown in Fig. 3. The shape of the talus from −3, −2, −1 deviations, mean shape to +1, +2, +3 deviations were repeated in 10 normal and 16 clubfoot models. The Std PC scores were calculated. The mean (± standard deviation) of the first five standardized principal component (Std PC) scores are shown in Table 2.

The mean differences between two groups were significant in first and fourth Std PC scores as shown in Fig. 4. The standard deviation of all Std PC scores

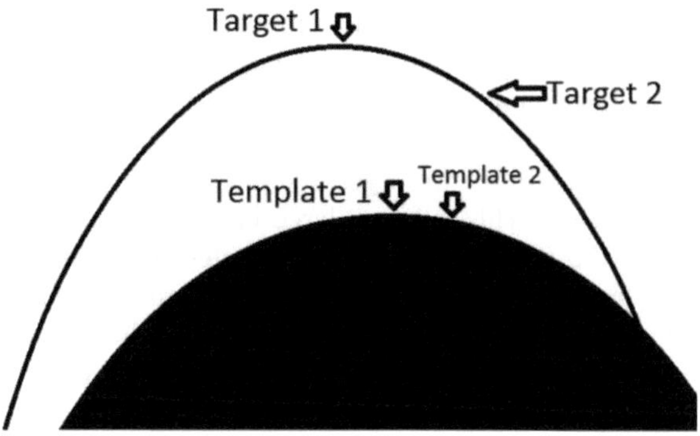

Fig. 5 Improvement of correspondence

except the first was larger in the clubfoot group. This indicated that SSM also found significant differences between the control and clubfoot groups.

Discussion

Direct linear measurements showed significant differences between the clubfoot and control groups in four distinct categories: width of the talus on the superior surface, angle of the talar neck, depth of the talar neck and talar volume. Clinical analysis of these differences would need to be performed to make definitive claims about the effects of these differences, but the short neck depth and steeper angle could be hindering dorsiflexion in clubfoot patients. Direct volume measurement results show similar trends to previously published work. Two studies from the Osaka School of Medicine found a 20.7% reduced talar volume in clubfoot patients (Itohara et al. 2005a, b), while this study found a 26% volume reduction. Our study did not analyse the effect of talonavicular alignment on clubfoot, but those results have been well documented in other work (Moritani et al. 2000; Saito et al. 2004; Kamegaya et al. 2001). However, our analysis of mortise width in comparison to talar width is a new approach to the geometric analysis of clubfoot. A full list of measurement results can be seen in Table 1.

In the SSM analysis, for the shape registration step, the selection of template was important. Since the morphology of some bones was particularly dissimilar to the others, the "Talus #3" bone was selected since it had a relatively similar shape to the mean model and it had the largest volume.

During the nonrigid shape registration process, the correspondence of vertices between the template and target shape was defined by checking each vertex from the template and finding its closest vertex in the target shape. This method would sometimes introduce large errors since some vertices from the template were aligned to the same point on the target shape and some other vertices on the target shape had no corresponding points on the template. For example, the black shape in Fig. 5 represents the aligned template and the white shape represents the target shape. If two points on the template, template 1 and template 2, and two points on the target shape, target 1 and target 2, were considered, a better correspondence would be found by aligning template 1 to target 1 and aligning template 2 to target 2. However, if the previous method was applied, both template 1 and template 2 would be aligned to target 2 since it was the closest vertex in the target shape to both points. This led to a large error because the information of the target shape around target 1 was missing. The algorithm was improved to avoid this situation. For the new alignment method, the correspondence was checked twice. The first time was as just described, finding the closest point in the target shape to each vertex from template as matrix A, i.e. both template 1 and 2 were aligned to target 2. Then the same algorithm was used in reverse to find the closest point in the template to each vertex from target shape as matrix B, i.e. target 1 was aligned to template 1. If the point like target 1 was not in A, it indicated that no point from the template was aligned to this point. Then this row of matrix A would be replaced by the corresponding row of matrix B, i.e. aligning template 1 to target 1 instead of target 2. This improvement greatly increased the quality of nonrigid shape registration.

Twenty-six models were used for SSM including only 10 normal shapes. For the future work, more data would need to be included for the test. Because there was a relatively larger variation in the paediatric talus, more models would improve statistical analysis. Also, the patients from the clubfoot group were the ones that retained residual deformity after treatment. These clubfoot bones were therefore a subset from paediatric clubfoot. For future work, another clubfoot group with talar bones from patients that did respond to treatment could be included.

In conclusion, both the direct geometric measurement and SSM methods found differences between control and clubfoot groups. Among 17 direct geometric measurements, 9 of them statistically different between groups. For SSM, the first 15 modes could explain about 90% of the variation. The mean Std PC scores were statistically different and the larger standard deviations in the clubfoot group indicated differences between two groups. Finally, an improved method for non-rigid shape registration was introduced that applies a two-way correspondence.

References

Allen B, Curless B, Popović Z (2003) The space of human body shapes: reconstruction and parameterization from range scans. ACM Trans Graph 22:587–594

Baek S-Y, Lee K (2012) Parametric human body shape modeling framework for human-centered product design. Comput Des 44:56–67

Bolkart T, Wuhrer S (2015) A groupwise multilinear correspondence optimization for 3d faces. Int J. Comput Vis. doi:10.1109/ICCV.2015.411

Gerig G, Styner M, Shenton ME, Lieberman JA (2001) Shape versus size : improved understanding of the morphology of brain structures. In: 4th international conference on medical image computing and computer-assisted intervention. LNCS 2208, pp 24–32. doi:10.1007/3-540-45468-3_4

Golland P, Grimson WEL, Shenton ME, Kikinis R (2005) Detection and analysis of statistical differences in anatomical shape. Med Image Anal 9:69–86

Itohara T et al (2005a) Assessment of talus deformity by three-dimensional MRI in congenital clubfoot. Eur J Radiol 53:78–83

Itohara T et al (2005b) Assessment of the three-dimensional relationship of the ossific nuclei and cartilaginous anlagen in congenital clubfoot by 3-D MRI. J Orthop Res 23:1160–1164

Kamegaya M, Shinohara Y, Kuniyoshi K, Moriya H (2001) MRI study of talonavicular alignment in club foot. J Bone Joint Surg Br 83:726–730

Moritani T et al (2000) MR evaluation of talonavicular angle in congenital talipes equinovarus. Clin Imaging 24:243–247

Patenaude B, Smith SM, Kennedy DN, Jenkinson M (2011) A Bayesian model of shape and appearance for subcortical brain segmentation. Neuroimage 56:907–922

Pekindil G, Aktas S, Saridogan K, Pekindil Y (2001) Magnetic resonance imaging in follow-up of treated clubfoot during childhood. Eur J Radiol 37:123–129

Rueckert D et al (1999) Nonrigid registration using free-form deformations: application to breast MR images. IEEE Trans Med Imaging 18:712–721

Saito S, Hatori M, Kokubun S, Abe Y, Kita A (2004) Evaluation of calcaneal malposition by magnetic resonance imaging in the infantile clubfoot. J Pediatr Orthop Part B Eur Paediatr Orthop Soc Pediatr Orthop Soc N Am 13:99–102

Smoger LM et al (2015) Statistical modeling to characterize relationships between knee anatomy and kinematics. J Orthop Res 33:1620–1630

Styner M et al (2006) Framework for the statistical shape analysis of brain structures using SPHARM-PDM. Insight J 1071:242–250

Wuhrer S, Shu C (2013) Estimating 3D human shapes from measurements. Mach Vis Appl 24:1133–1147

Growth and Remodeling of Tissue Structure and Properties

Yoram Lanir

Abstract Soft biological tissues vary significantly in their mechanical properties, from the more rigid articular cartilage to the very soft and extensible skin and mesentery. Yet tissues are made of the same constituents: Fibers, muscle cells, non-muscle cells, and a fluid matrix. The key to diversity in properties is a parallel diversity in structure. The present study addressed the question of how is tissue structure determined? Living tissues have the unique ability to grow and remodel under altered mechanical loading by turnover of their fibers, where some are degraded and new ones are produced and deposited. It was hypothesized that tissue structure evolves with growth by remodeling its structure in response to growth-induced loading. The hypothesis was tested by structural simulation. The modeling framework developed is a multi-scale, micro-mechanical one, which integrates the effects of cells, fibers, and matrix, based solely on the biological processes in the remodeling tissue, thereby linking the constituents' turnover to the evolving tissue structure and properties. The results are compatible with the evolved adult tissue structure and mechanical characteristics. Specifically, the theory predicts the evolution of well-known soft tissues features such as the nonuniform undulation of collagen fibers and associated tissue, nonlinear convex strain–stress response, and the evolution of growth-induced prestrain and prestress. These results support the notion that tissues' structure and properties evolve as they grow.

Introduction

Soft tissues vary significantly in their mechanical properties. Yet they are made of the same constituents: fibers (collagen, elastin, micro-fibrils), muscle cells (skeletal, myocytes, smooth muscle cells), non-muscle cells (e.g., fibrocytes), and

Y. Lanir (✉)
Faculty of Biomedical Engineering, Technion - Israel Institute of Technology, Haifa, Israel
e-mail: yoramlanir@yahoo.com

A. Gefen and D. Weihs (eds.), *Computer Methods in Biomechanics and Biomedical Engineering*, Lecture Notes in Bioengineering,
DOI 10.1007/978-3-319-59764-5_30

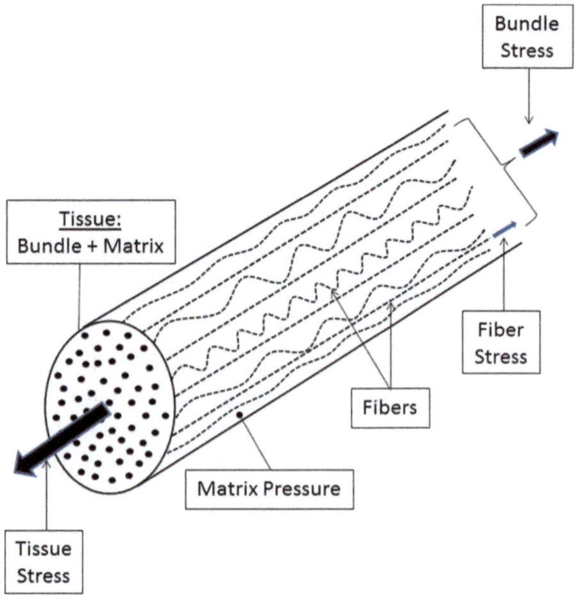

Fig. 1 A structural scheme of the simulated uni-directional tissue in which the parallel collagen fibers are embedded in a fluid-like matrix. From Lanir (2015)

extra-cellular matrix (water, ions, soluble charged proteoglycans). Structure, which is tissue specific, is the key to the diversity of properties.

Some tissues like, tendon and ligament, consist of uni-directional single-type collagen fibers (Fig. 1). Other tissues, like the skin and mesentery, are flat and consist of both collagen and elastin fibers.

More complex structures are those of blood vessels and the myocardium which consist of three-dimensional (3D) networks of passive collagen and elastin fibers, and of contractile smooth muscle cells and myocytes, respectively. Regardless of the diversity in consistency and organ geometry, common features observed in most soft tissues are dispersion in their fiber orientation (in multidimensional tissues) and undulation of their fibers when unloaded. Both are tissue specific and are nonuniform among fibers of each tissue. These two structural attributes (orientation and undulation dispersions) bear significantly on the tissue mechanical properties.

This study addressed the question of how tissue structure is determined. More specifically, is structure genetically determined, or does it evolve with growth? It is hypothesized that tissue structure evolves with growth. This hypothesis was tested by model simulation. As a first step aimed to demonstrate the feasibility of this hypothesis, the study considered the case of uni-directional tissues consisting of collagen fibers (Fig. 1).

Living tissues are unique amongst all materials in their ability to adapt to altered mechanical environment in a process termed growth and remodeling (G&R). At adulthood, G&R is manifested under change in physical activity (e.g., exercise, immobilization), during pregnancy, and as a result of pathologies (e.g., abdominal and cerebral aneurisms, hypertension). A large body of evidence is consistent with notion that G&R is cell-induced: the extra-cellular matrix (ECM) is remodeled by the cells to maintain homeostatic mechanical environment needed by the cells to survive and proliferate. ECM remodeling occurs by fibers turnover, where some fibers are degraded in a loading-dependent rate, while others are synthesized and deposited under stretch on extant fibers.

Collagen is enzymatically degraded by matrix metalloproteinases (MMPs) which are synthesized by the ECM cells. The susceptibility to the enzyme is determined by the collagen conformational freedom. In fibrils, the collagen molecules are closely packed, have limited freedom to move under thermal agitation, and are thus more resistant to the enzyme binding. Under high stretch however, the rate of proteolysis increases by at least two mechanisms. First, stretch disrupts the fibril lattice thereby increasing the collagen freedom to move and with it, its susceptibility to the enzyme. In addition, it has been suggested (Willett et al. 2007) that stretch exposes susceptible cryptic sites along the collagen such as its thermally labile domain (TLD). Collagen fibrillogenesis occurs in the ECM cells (primarily fibrocytes and SMCs). Collagen monomers are synthesized in intra-cellular vacuoles. The fibril is then created from these monomers when the vacuoles align and fuse laterally to form long fibril-forming compartments where sub-fibrils are polymerized. These compartments then fuse further laterally into bundle forming compartments where thicker fibrils are formed (Birk and Trelstad 1986).

Following synthesis, the collagen fibrils are deposited on extant fibers (Birk et al. 1989; Nimni 1990) under homeostatic stretch which is tissue specific (Eastwood et al. 1996; Harris et al. 1981).

The three G&R steps (degradation, synthesis and deposition under stretch) were structurally simulated in this study based solely on the biology of the fibers turnover, linking it to the evolving tissue structure.

Materials and Methods

The simulation was directed by data-based assumptions. The main ones are:

- The tissue is composed of fibers, cells and a fluid-like matrix.
- There are no voids, and all constituents are incompressible.
- Under stretch, the fibers are linear elastic. When contracted, the thin and long fibers buckle under insignificantly low force.
- The matrix is under hydrostatic pressure induced by the osmotic activity of its charged proteoglycan macro-molecules.
- Synthesized fibrils are deposited on extant fibers under stretch which is normally distributed within the homeostatic range.
- The total tissue stress is the sum of its constituents' contributions, i.e., that of the fibers stress and the matrix pressure (the "rule of mixtures").
- The rates of both degradation and deposition of fibers have both stretch-dependent and basal contributions.
- The fibers degradation and deposition are first-order reactions.

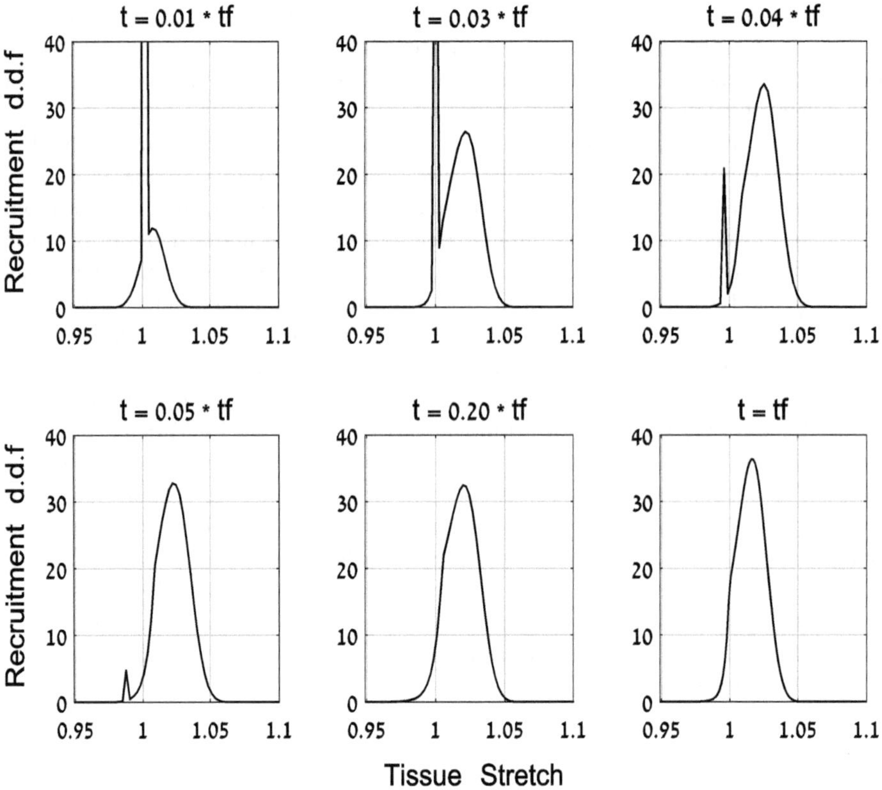

Fig. 2 Evolution of the fibers recruitment stretch density distribution. Fibers with recruitment stretch lower than unity are already stretched in the unloaded state under the effect of the matrix pressure. "tf" is the total growth time. From Lanir (2015)

Details of the numerical simulation, of the model equations, and of the selected levels of the data-based parameters, are given elsewhere (Lanir 2015). Briefly, the growth protocol imposed on the tissue is a slow constant stretch rate of 0.50 per year for a period of two years. Three aspects of the evolving tissue properties were specifically addressed: the distribution of the fibers undulation, indexed experimentally by their recruitment stretch distribution; the associated evolved tissue mechanical response to short stretch protocol; and the onset of growth-induced tissue pre-stress and pre-stretch. Initially, in the stress-free state, all fibers were taken to be straight and unstretched.

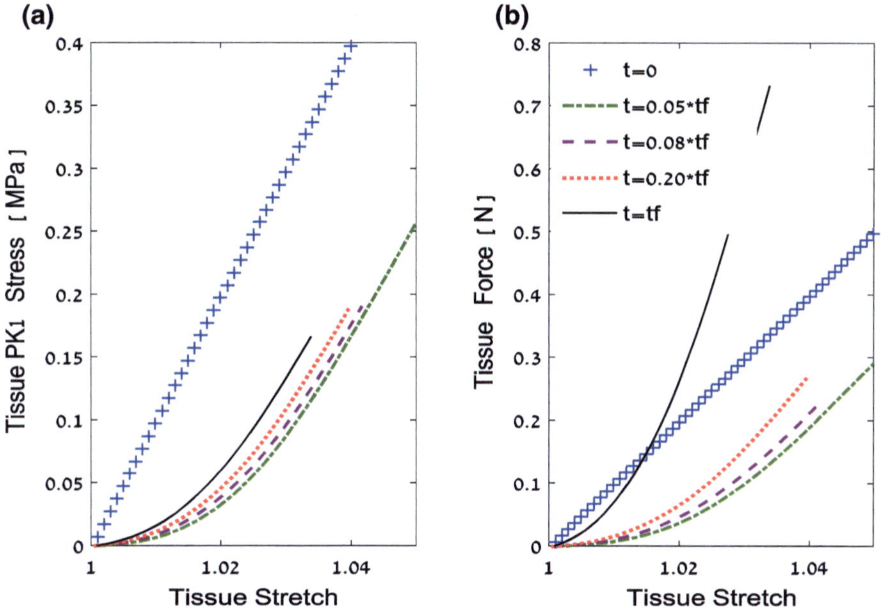

Fig. 3 Evolution of the tissue response to identical short stretch protocol. **a** The tissue first Piola–Kirchoff stress. **b** The total tissue force which incorporates the increase of cross-sectional area with growth. Adapted from Lanir (2015)

Results

The predicted evolution of the tissue structure in terms of the density distribution of the fibers recruitment stretch is depicted in Fig. 2. It evolves rapidly with growth time from a uniform value of one for all fibers into a bell-shaped distribution. The signature of the initial uniform distribution fads away already after 5% of the growth time (the sub-plot titled "t = 0.05 * tf" in Fig. 2). In fact applying a different and nonuniform initial distribution of the fibers, recruitment stretch was found to have no effect on the evolved structure after 10% of the growth time.

The predicted functional consequences of the tissue evolving structure are clearly manifested in its mechanical response to identical short test protocols along the growth period (Fig. 3). The initial linear response of the uniformly straight fibers develops gradually into convexly nonlinear one.

The simulation predicts that growth induces onset of in situ pre-stress and associated pre-stretch. Their evolutions are presented in Fig. 4. They seem to asymptotically approach constant levels.

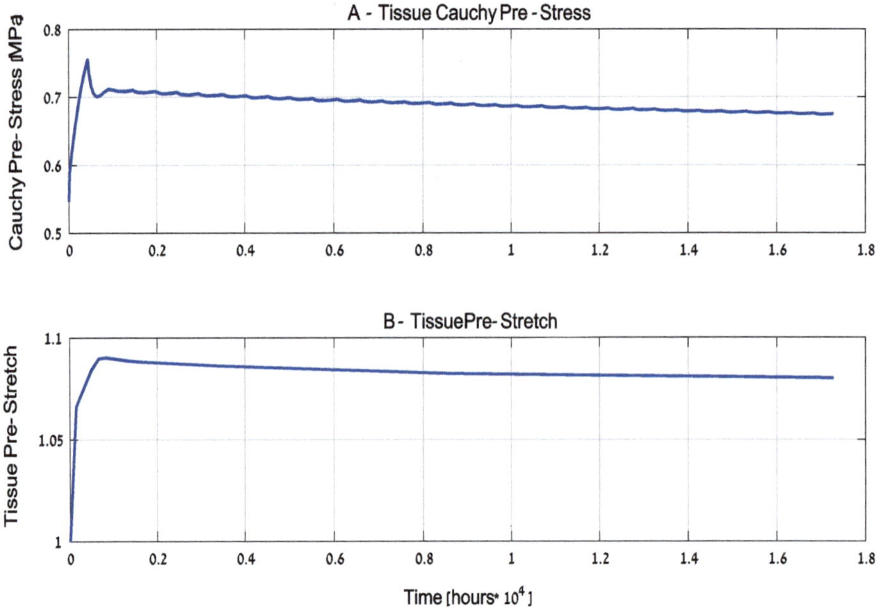

Fig. 4 Growth-induced onset of in situ pre-stress (**a**) and tissue pre-stretch (**b**). Adapted from Lanir (2015)

Discussion

Predictions of the mechano-biological G&R simulations are compatible with commonly observed features of evolved structure and mechanics of soft tissues, specifically in terms of the bell-shaped distribution of the fibers recruitment (Fig. 2), the associated convex nonlinear mechanical response (Fig. 3), and the onset of pre-stress and pre-strain (Fig. 4). These results are consistent with the hypothesis that tissue structure and mechanics evolve with growth by remodeling of its constituents.

These conclusions are expected to be insensitive to the constitutive assumptions adopted. Specifically, the additive rule of mixtures guarantees that nonlinear, inelastic, and contractile fibers can be incorporated as well without affecting the response of each other. In addition, non-first-order kinetics of the turnover reactions could change the rate of evolution of the tissue features, but not their evolved mature characteristics.

References

Birk DE, Trelstad RL (1986) Extracellular compartments in tendon morphogenesis: collagen fibril, bundle, and macroaggregate formation. J Cell Biol 103(1):231–240

Birk DE, Southern JF, Zycband EI, Fallon JT, Trelstad RL (1989) Collagen fibril bundles: a branching assembly unit in tendon morphogenesis. Development 107(3):437–443

Eastwood M, Porter R, Khan U, McGrouther G, Brown R (1996) Quantitative analysis of collagen gel contractile forces generated by dermal fibroblasts and the relationship to cell morphology. J Cell Physiol 166(1):33–42

Harris AK, Stopak D, Wild P (1981) Fibroblast traction as a mechanism for collagen morphogenesis. Nature 290(5803):249–251

Lanir Y (2015) Mechanistic micro-structural theory of soft tissues growth and remodeling: tissues with unidirectional fibers. Biomech Model Mechanobiol 14(2):245–266

Nimni ME (1990) Collagen in cardiovascular tissues. In: Hastings GW (ed) Cardiovascular biomaterials. Springer, New York

Willett TL, Labow RS, Avery NC, Lee JM (2007) Increased proteolysis of collagen in an in vitro tensile overload tendon model. Ann Biomed Eng 35(11):1961–1972

Proximity of Metastatic Cells Strengthens the Mechanical Interaction with Their Environment

Yulia Merkher and Daphne Weihs

Abstract The main cause of cancer-related deaths is metastasis-spreading of cancer cells to different sites in the body. A critical step in metastasis formation is invasion of cells through the surrounding tissue. During invasion, cancer cells change their shape and apply forces. We have previously identified that about 30% of single, metastatic, breast cancer cells will indent impenetrable synthetic, non-degradable, polyacrylamide gels, when gel stiffness is in the range 1–10 kPa. By measuring the depth of indentation of an initially flat gel and monitoring time-dependent microscopic changes in cell morphology, we were able to distinguish between benign and metastatic cells, also identifying their metastatic potential (MP); benign cells do not indent the gels. Recent works have indicated that metastases from solid tumors occur predominantly by collective cell invasion. Hence, in the current study we evaluate the mechanical interactions of cell clusters with the impenetrable gel. We observe that indenting subpopulations of metastatic cells are doubled in clusters, and cells are also indent more deeply; this increases likelihood to successfully form metastasis in the body. Concurrently, double the fraction of high MP cells indent gels as compared to low MP cells, while benign cells do not indent even in clusters. We also show that the gel platform can be used to determine the time-dependent impact of chemotherapeutics on the cells' ability to apply forces and indent gels. Our approach can provide a rapid, mechanical prediction of the likelihood for invasiveness of cancer cells and can further be applied in a patient-specific approach, thus providing a personalized prognosis that may improve treatment of cancer patients and increase their life expectancy.

Y. Merkher · D. Weihs (✉)
Faculty of Biomedical Engineering, Technion - Israel Institute of Technology, 3200003 Haifa, Israel
e-mail: daphnew@technion.ac.il

A. Gefen and D. Weihs (eds.), *Computer Methods in Biomechanics and Biomedical Engineering*, Lecture Notes in Bioengineering,
DOI 10.1007/978-3-319-59764-5_31

Introduction

The main cause of cancer-associated mortality is metastasis-spreading of cancer cells to different sites in the body; this accounts for 90% of deaths. A critical step in metastasis formation is invasion of cells through the surrounding tissue, including cells and the extracellular matrix. During invasion, cancer cells change their shape and apply forces. We have previously identified that about 30% of single, high metastatic potential (MP), breast cancer cells will indent impenetrable synthetic, non-degradable, polyacrylamide gels, when gel stiffness is in the range 1–10 kPa (Kristal-Muscal et al. 2013, 2015). By measuring the depth of indentation of an initially flat gel and monitoring time-dependent microscopic changes in cell morphology, we were able to distinguish between benign and metastatic cells, also to identify their metastatic potential (Kristal-Muscal et al. 2013, 2015; Dvir et al. 2015); benign cells do not indent the gels.

Depending on the cell type and tissue environment, cells can migrate and invade in two main ways: individually or collectively, as multicellular groups. Most invasive solid tumors display predominantly collective invasion (Cheung et al. 2013), in which groups of cells invade their surroundings while cells remain either physically or chemically connected (Friedl et al. 2012). Hence, in the current study we evaluate the mechanical interactions of cell clusters with an impenetrable gel.

Materials and Methods

Cell Culture

We have used three commercially available human, epithelial breast cell lines: high MP cells-MDA-MB-231, low MP cells-MDA-MB-468, benign cells-MCF-10A (all cells from ATCC, Manassas, VA). Both metastatic cell lines were collected from lung metastases. The cells were cultured in 37 °C, 5% CO_2, 90% humidity and in their appropriate media: Dulbecco's modified Eagle's medium (DMEM) (Gibco, Invitrogen life technologies, CA) with supplementation of 10 vol.% fetal bovine serum (Hyclone, ThermoFisher Scientific, Waltham, MA), 1 vol.% each of L-glutamine, penicillin-streptomycin and sodium pyruvate (all from Biological Industries, Kibbutz Beit Haemek, Israel) for metastatic cells or with supplementation of 5 vol.% horse serum (Hyclone, ThermoFisher Scientific, Waltham, MA), 0.05 vol.% Hydrocortisone, 0.01 vol.% Cholera toxin, 0.1 vol.% Insulin (all from Sigma, St Louis, MO), 1 vol.% penicillin-streptomycin, 1 vol.% L-Glutamine (both from Biological Industries, Israel) and 0.01 vol.% Human EGF (Peprotech Asia, Israel) for benign cells.

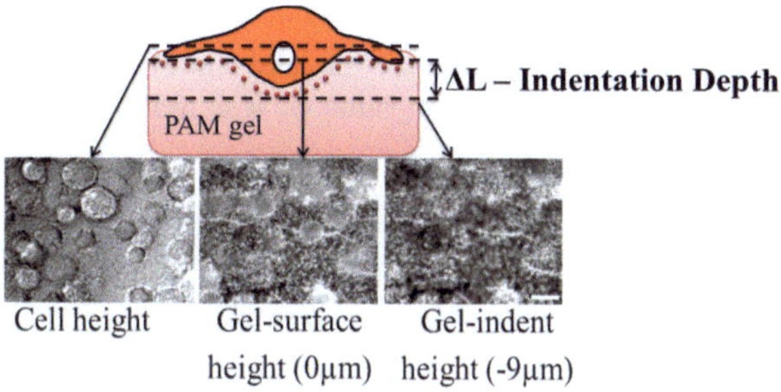

Fig. 1 Indentation depth measurement. The highly metastatic potential cells seeded on 2.4 kPa PAM gel pushing the fluorescent beads imbedded at gel surface into the gel. The indentation depth is measured from difference in focal planes of beads at gel surface and beads pushed in-depth under the indenting cells. *Scale bar* is 20 μm

Hydrogel Preparation

We have used polyacrylamide (PAM) hydrogel with fluorescent beads (200 nm) imbedded at its surface, prepared according to an established protocol (Kristal-Muscal et al. 2013, 2015). The gels were prepared at stiffness of 2400 Pa which is in a range of physiological stiffnesses of breast and lung (Levental et al. 2007).

Imaging and Indentation Depth Calculation

We seeded on separate gels 300,000 of each cell type, within their respective media. Approximately 45 min after seeding we began imaging the gels with the cells using an inverted, epifluorescence Olympus IX81 microscope, equipped with a 60x/0.7NA differential interference contrast (DIC, Nomarsky optics) air-immersion, long working-distance objective lens. At each randomly chosen field-of-view on the gel (typically 10 fields per gel), at least three images were taken: (1) a DIC image of the cells on the gel, (2) a fluorescence image of the particles embedded at the focal plane of the gel surface, and (3) a series of fluorescence images at the lowest focal depth where particles were in focus, identifying the indentation depth of each cell (Fig. 1). The indentation depth was calculated from difference in focal planes of fluorescent images. We also determined the number of indenting cells out of the total cells.

Statistical Analysis

Statistical analysis was performed to compare between the different cells types or to compare single and multiple cells of the same type. For comparison between the means of two groups, the general, linear mixed model, which is a multivariate regression method that helps to generalize the analysis of variance (ANOVA) was used (Edwards 2000). In all cases, a p value of less than 0.05 is considered statistically significant.

Results

We evaluated the number of indenting cells of each cell within the group, comparing high metastatic potential, low MP, and benign cells. We observe that the percentage of indenting subpopulations of metastatic cells are doubled in clusters, while benign cells almost did not indent the gels regardless of seeding density (Fig. 2). Concurrently, double the fraction of high MP cells indent gels as compared to low MP cells, while benign cells do not indent even in the 2-dimensional clusters. That could be explained by higher aggressiveness of cancer clusters compared to single cancer cell (Clark and Vignjevic 2015). Amplified invasiveness of cancer cell clusters was shown decades ago, when mice injected with 3-dimensional clumps of tumor cells had developed a higher number of metastases when compared to mice injected with an equal total number of single cancer cells (Fidler 1973).

The adjacent cells also indent more deeply. While single cells display Gaussian-like distribution of indentation depths (not shown) with no cells indenting deeper than 10 μm, the multicellular groups display a wide, bi-modal-like distribution with appearance of second indentation depth peak (Fig. 3). The indentation depths of single cells with high and low MP did not significantly differ. However, when cells are adjacent and in group organization, most of the high MP cells indented more deeply, in the second indentation peak. In contrast, most of the low MP cells still indented similarly to the single cells. Therefore we could distinguish between the cells with the high and low MP based on the number of deeply

Fig. 2 The percentage of single (*light colors*) and multiple (*dark colors*) cells with high and low MP and benign cells used as control, indenting into synthetic impenetrable 2.4 kPa gel

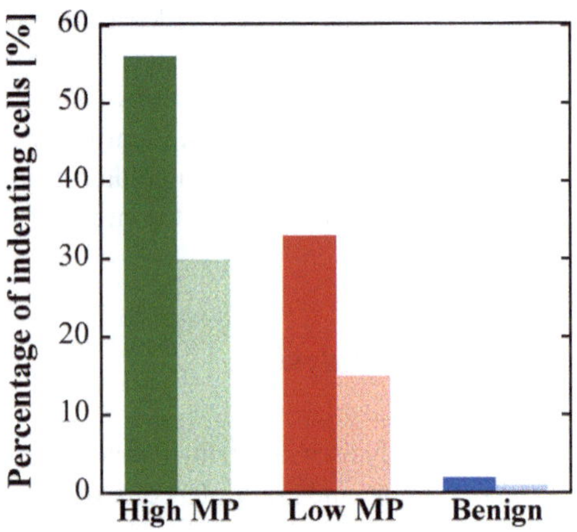

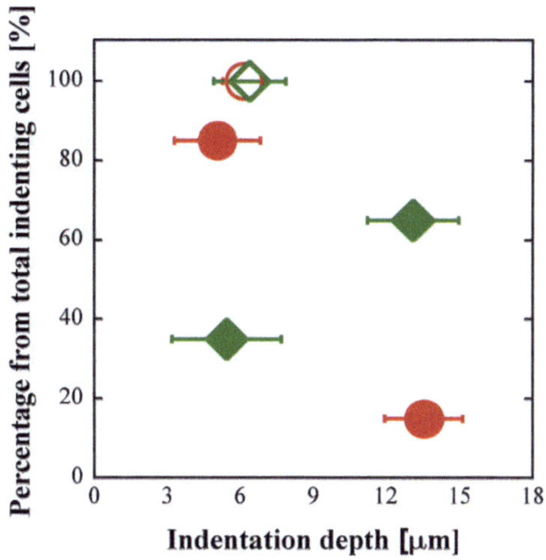

Fig. 3 The percentage (from total indenting cells) of high (*green diamonds*) and low (*red circles*) MP cancer cells indenting to average indentation depths (*error bars* are standard deviations). All single cells (*empty markers*) not indenting deeper than 10 μm. Most of adjacent cells with high MP indenting to the second indentation peak while most of the adjacent cells with low MP indenting to the first peak

indenting cells. A similar phenomenon was obtained with intravital multiphoton microscopy for in vivo migration of cell clusters, while the velocity of single highly invasive (MDA-MB-231) and invasive (TN1) breast cancer cells was not significantly different, the multicellular groups of highly invasive cells enhance their velocity significantly (Patsialou et al. 2013).

Discussion

We found evidence that the groups of metastatic breast cancer cells from cell lines are able to indent in larger numbers and more deeply than single cells. That likely indicates that cell-groups are more aggressive/invasive than single cells and the proximity of metastatic cells increases likelihood to successfully form metastasis in the body. Using this approach we may be able to provide a rapid, mechanical prediction of the likelihood for invasiveness of cancer cells. Definitely, before our approach could be used as metastasis prediction platform a lot of future work with various cell lines and cells resected from primary tumors should be done. This approach can be potentially applied in a patient-specific way, thus providing a personalized prognosis that may improve treatment of cancer patients and increase their life expectancy.

Acknowledgements The work was partially supported by The Technion EVPR Funds-The Elias Fund for Medical Research and The Karbeling Fund for Bio-Medical Engineering Research, and also by a grant from the Ministry of Science, Technology and Space, Israel, and the National Science Council (NSC) of Taiwan.

References

Cheung KJ, Gabrielson E, Werb Z, Ewald AJ (2013) Collective invasion in breast cancer requires a conserved basal epithelial program. Cell 155(7):1639–1651

Clark AG, Vignjevic DM (2015) Modes of cancer cell invasion and the role of the microenvironment. Curr Opin Cell Biol 36:13–22

Dvir L, Nissim R, Alvarez-Elizondo MB, Weihs D (2015) Quantitative measures to reveal coordinated cytoskeleton-nucleus reorganization during in vitro invasion of cancer cells. New J Phys 17(4):43010

Edwards LJ (2000) Modern statistical techniques for the analysis of longitudinal data in biomedical research. Pediatr Pulmonol 30(4):330–344

Fidler IJ (1973) The relationship of embolic homogeneity, number, size and viability to the incidence of experimental metastasis. Eur J Cancer 9(3):223–227

Friedl P, Locker J, Sahai E, Segall JE (2012) Classifying collective cancer cell invasion. Nat Cell Biol 14(8):777–783

Kristal-Muscal R, Dvir L, Weihs D (2013) Metastatic cancer cells tenaciously indent impenetrable, soft substrates. New J Phys 15:35022

Kristal-Muscal R, Dvir L, Schvartzer M, Weihs D (2015) Mechanical interaction of metastatic cancer cells with a soft gel. Procedia IUTAM 12:211–219

Levental I, Georges PC, Janmey PA (2007) Soft biological materials and their impact on cell function. Soft Matter 3(3):299–306

Patsialou A, Bravo-Cordero JJ, Wang Y, Entenberg D, Liu H, Clarke M, Condeelis JS (2013) Intravital multiphoton imaging reveals multicellular streaming as a crucial component of in vivo cell migration in human breast tumors. Intravital 2(2):e25294

Patient-Specific Numerical Model of Calcific Aortic Stenosis and Its Treatment by Balloon-Expandable Transcatheter Aortic Valve: Effect of Positioning on the Anchorage

Gil Marom, Matteo Bianchi, Ram P. Ghosh and Danny Bluestein

Abstract Despite the promising outcomes of transcatheter aortic valve replacement (TAVR), adverse events may occur as a result of suboptimal placement. The aim of this study is to evaluate the effect of various TAVR deployment positions on the risk for intra-procedural migration and of post-procedural paravalvular regurgitation. Finite Element (FE) and Computational Fluid Dynamics (CFD) models are presented for several procedural scenarios for a patient-specific morphology. Crimping and deployment of balloon-expandable Edwards SAPIEN were modeled in three locations. The proximal deployment resulted in higher risk for migration, while the distal and midway positioning resulted in comparable outcomes. These resulting configurations were used to assess diastolic hemodynamics. The distal case had preferred hemodynamics compared to the midway with more limited leakage. The proposed approach has the potential to be used in procedural planning to ultimately achieve better clinical outcomes.

Introduction

Calcific aortic stenosis is a degenerative process in which the aortic valve narrows due to the formation and growth of calcium deposits on its leaflets. Transcatheter aortic valve replacement (TAVR) has become the only lifesaving solution for patients who cannot undergo the standard surgical valve replacement. Despite its promising outcomes, adverse events such prosthesis migration have been reported (Cao et al. 2012) and peri-procedural complications such as paravalvular leakage (PVL) and cardiac conduction abnormalities may occur as a result of suboptimal placement, leading to poor device performance. The impact of these complications has, therefore, hindered TAVR expansion to lower risk patients, even though

G. Marom (✉) · M. Bianchi · R.P. Ghosh · D. Bluestein
Biomedical Engineering, Stony Brook University, Stony Brook, NY, USA
e-mail: gil.marom@stonybrook.edu

A. Gefen and D. Weihs (eds.), *Computer Methods in Biomechanics and Biomedical Engineering*, Lecture Notes in Bioengineering,
DOI 10.1007/978-3-319-59764-5_32

comparable outcome to surgical treatment was reported in terms of survival rate and stroke prevalence (Leon et al. 2016). The aim of this study is to evaluate the effect of various TAVR deployment positions on the procedural outcome by assessing the risk for intra-procedural migration and of post-procedural paravalvular regurgitation. This was achieved by building Finite Element (FE) and Computational Fluid Dynamics (CFD) models capable of replicating several procedural scenarios for a patient-specific morphology.

Materials and Methods

The first stage was to model the balloon-expandable Edwards SAPIEN valve crimping process. The balloon deflation and the stent crimping were modeled independently in the FE solver Abaqus 6.14 Explicit. Subsequently, the crimped stent was deployed by balloon-inflation in a patient-specific calcified aortic root, reconstructed from pre-interventional CT scans of a patient that endured valve migration into the left ventricle (LV) during his TAVR procedure in Stony Brook University Hospital. Additional details on the reconstruction of the patient-specific anatomy and the FE analyses of the TAVR stent and balloon are given in our recent study (Bianchi et al. 2016).

The deployment location was parametrized in three configurations: midway, in which stent centroid is placed on the annulus plane, distal, and proximal, where the stent centroid was shifted by 30% towards the aorta and the LV, respectively. The anchorage of the stent to the native tissue was quantitatively assessed based on the contact between the stent and the root during the deployment and the recoil phases.

The resulting deformed configurations were used to generate new models to assess post procedural hemodynamics during diastole. Particularly, the PVLs were investigated and quantified. The walls of the fluid domain, from both the vessel and the prosthetic stent, were processed in ANSYS 17 SpaceClaim to obtain cleaned and merged domains while retaining the detailed features. The prosthetic leaflets and cuff were deformed from their pre-crimped to the end-recoil configuration by connecting them to the Edwards SAPIEN stent and displacing the nodes of the stent. Then the valve was closed by applying diastolic transvalvular pressure difference on the aortic side of the prosthetic leaflets. The resulting valve geometry was smoothed and assembled with the root to create the fluid domain and was meshed in ANSYS Fluent Meshing 17 with tetrahedral cells. The aortic inlet and the coronary outlets were extruded with a length of 5 hydraulic diameters in order to minimize the effect the boundary conditions (BCs) on the region of interest. The funnel-shaped ventricular outflow tract was created by extruding and scaling the aortic annulus edge. The mesh was converted to polyhedra and transient diastolic flow analyses were conducted in ANSYS Fluent 17, assuming homogeneous and Newtonian blood and laminar flow. The aortic and the ventricular pressure waveforms and the main coronary arteries flow waveforms were applied as BCs. Additionally, approximately 30,000 particles were injected in the fluid domain

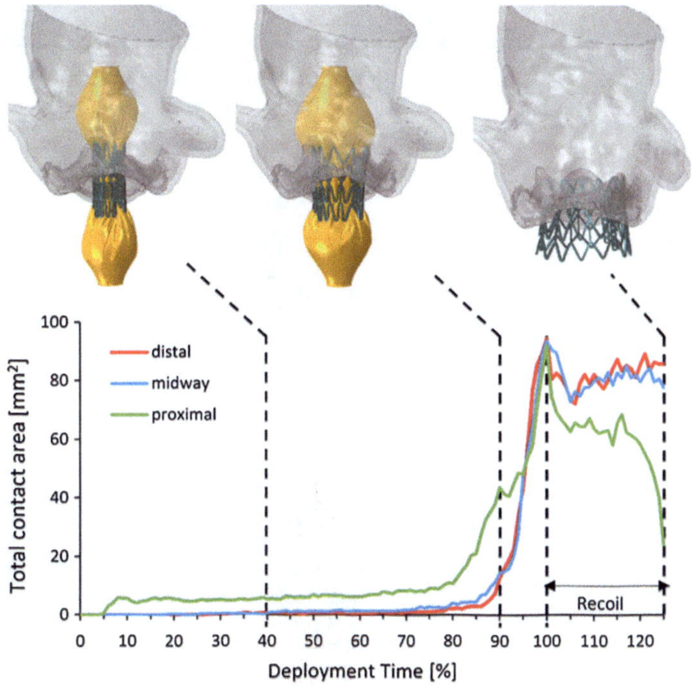

Fig. 1 Total contact area between the stent and the native valve calculated during the deployment (up to 100%) and the recoil phase (100–120%) for the three configurations. At the top, snapshots of the deployment of the stent (teal) in the midway configuration in several instances: at the 40% of the deployment time ("dogbone" balloon configuration, *left*), at 90% (rapid stent expansion, *middle*) and at the end of the recoil (*right*). Adapted from Bianchi et al. (2016) with permission

above the stent at the beginning of the simulation in order to track their trajectories through the paravalvular gaps for each configuration. The root mean square of the regurgitating flow rate was calculated and the overall area of the paravalvular gaps was estimated through effective orifice area (EOA).

Results

The proximal deployment resulted in higher risk for valve migration, with a 70% decrease in total contact area at the beginning of the recoil, when the stent migrated into the LV cavity (Fig. 1). The contact was observed to be highly localized leading to higher stress magnitudes in the native leaflets than in the other two configurations. The distal and midway positioning resulted in comparable outcomes in terms of contact area and stresses, even though the distal deployment provided slightly better anchorage. Given the migration found in the proximal configuration, the PVL was therefore assessed in the midway and distal cases only.

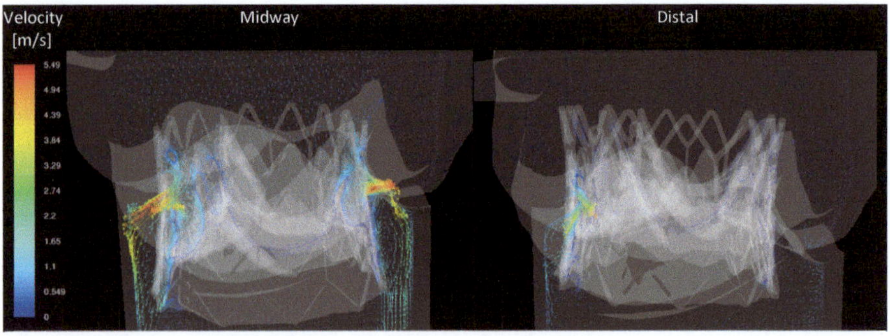

Fig. 2 Velocity vectors showing the paravalvular regurgitation in the midway and distal configurations at peak diastole

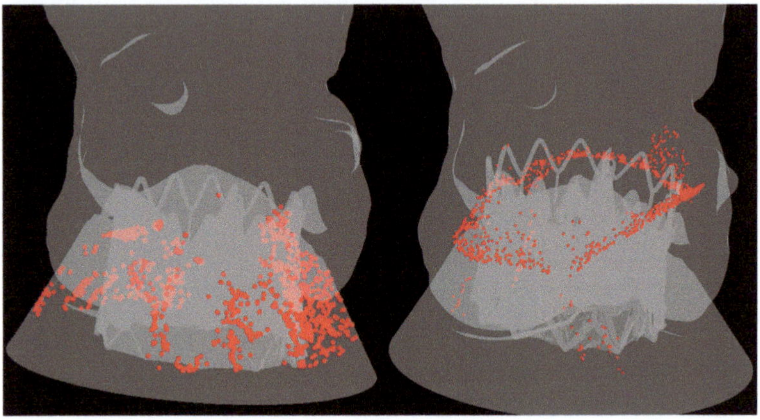

Fig. 3 Particles location at 0.4 s after the onset of diastole in the midway (*left*) and distal configuration (*right*). Most of the particles in the former case already passed through the paravalvular gaps, whereas in the latter case a large amount of particles remains in the sinus

Peak velocities were comparable (6.57 vs. 6.29 m/s for the midway and distal cases, respectively) and occurred in the instant when the aortic-ventricular pressure difference was the highest (100 ms after the onset of diastole). The midway case resulted in larger regurgitation regions in which the blood flowed through the stent struts, suggesting that the location of the stent is too proximal. The distal case had more limited PVL and confined in the region between the stent and the non-coronary leaflet (Fig. 2). Particle trajectories and velocity field comparison showed the presence of regurgitation in both configurations, with the midway case experiencing significantly higher flow through the struts. The distal case resulted in sixfold decrease in flow rate (32 vs. 207 ml/s) through the paravalvular gaps and comparable increase in EOAs (0.33 vs. 0.05 cm^2). Particles flowed through the gaps more rapidly in the midway configuration than in the distal one, confirming the existence of a more substantial leakage in the regions around the deployed stent (Fig. 3).

Discussion

The FE models accurately mimic the pathophysiological features of the patient's anatomy and simulate various procedural scenarios, thus providing quantifiable information about the anchorage of the TAVR stent to the native root. The CFD models were capable of investigating the post-deployment hemodynamics, specifically in the flow through the paravalvular gaps, showing a direct effect of the initial TAVR positioning on the degree of post-TAVR PVL. The proposed approach might be used as a predictive tool for procedural planning in order to ultimately achieve better clinical outcomes.

References

Bianchi M, Marom G, Ghosh RP, Fernandez HA, Taylor JR Jr, Slepian MJ, Bluestein D (2016) Effect of balloon-expandable transcatheter aortic valve replacement positioning: a patient-specific numerical model. Artif Org 40(12):E292–E304. doi:10.1111/aor.12806

Cao C, Ang SC, Vallely MP, Ng M, Adams M, Wilson M (2012) Migration of the transcatheter valve into the left ventricle. Ann Cardiothorac Surg 1(2):243–244. doi:10.3978/j.issn.2225-319X.2012.07.11

Leon MB, Smith CR, Mack MJ, Makkar RR, Svensson LG, Kodali SK, Thourani VH, Tuzcu EM, Miller DC, Herrmann HC, Doshi D, Cohen DJ, Pichard AD, Kapadia S, Dewey T, Babaliaros V, Szeto WY, Williams MR, Kereiakes D, Zajarias A, Greason KL, Whisenant BK, Hodson RW, Moses JW, Trento A, Brown DL, Fearon WF, Pibarot P, Hahn RT, Jaber WA, Anderson WN, Alu MC, Webb JG (2016) Transcatheter or surgical aortic-valve replacement in intermediate-risk patients. N Engl J Med. doi:10.1056/NEJMoa1514616

MIX
Papier aus verantwortungsvollen Quellen
Paper from responsible sources
FSC® C105338

If you have any concerns about our products,
you can contact us on
ProductSafety@springernature.com

In case Publisher is established outside the EU,
the EU authorized representative is:
Springer Nature Customer Service Center GmbH
Europaplatz 3, 69115 Heidelberg, Germany

Printed by Libri Plureos GmbH
in Hamburg, Germany